11th Edoardo Amaldi Conference on Gravitational Waves (AMALDI 11)

Journal of Physics: Conference Series
Volume 716

Gwangju, South Korea
21 – 26 June 2015

Editors:

Hyung Mok Lee
John Oh

ISBN: 978-1-5108-2454-6
ISSN: 1742-6588

Printed from e-media with permission by:

Curran Associates, Inc.
57 Morehouse Lane
Red Hook, NY 12571

Some format issues inherent in the e-media version may also appear in this print version.

Printed by Curran Associates, Inc. (2016)

For permission requests, please contact the Institute of Physics
at the address below.

Institute of Physics
Dirac House, Temple Back
Bristol BS1 6BE UK

Phone: 44 1 17 929 7481
Fax: 44 1 17 920 0979

techtracking@iop.org

Additional copies of this publication are available from:

Curran Associates, Inc.
57 Morehouse Lane
Red Hook, NY 12571 USA
Phone: 845-758-0400
Fax: 845-758-2633
Email: curran@proceedings.com
Web: www.proceedings.com

Table of contents

Volume 716

11th Edoardo Amaldi Conference on Gravitational Waves (AMALDI 11)
21–26 June 2015, Gwangju, South Korea

Accepted papers received: 29 April 2016
Published online: 2 June 2016

Preface

Papers

Preface

The 11th Amaldi Conference on gravitational waves was held from June 21 to 26, 2016 in Gwangju, Korea. This meeting was sponsored by the International Union of Pure and Applied Physics (IUPAP), Korean Astronomical Society, Korean Physical Society, Korean Federation of Science and Technology (KOFST), National Research Foundation (NRF), Korea Tourism Organism, and Gwangju Convention Bureau.

More than 150 scientists from 26 countries attended the 11th Amaldi Confernece. Although there was an outbreak of Middle East Respiratory Syndrome (MERS) just before the conference, there were very little cancellations. There was no incident related to the MERS disease.

The scientific program, including the list of the invited plenary speakers, was arranged by the Gravitational Wave International Committee (GWIC) chaired by Eugenio Coccia. Most of the communication between the GWIC and the Local Organizing Committee (LOC) was handled by Stan Whitcomb, general secretary of the GWIC. There were 10 plenary sessions and 13 parallel sessions. Two plenary sessions were held each morning from Monday to Friday while the afternoon was devoted to parallel sessions. A special lecture for the participants of the meeting on ultra-high intensity lasers was delivered by Chang Hee Nam, director of the Center for Relativistic Laser Science located at the Gwangju Institute of Science and Technology. Bernard Schutz gave a public talk entitled "Listening to Black Holes and the Big Bang - the Oldest Concert of the Universe" in the evening on June 23. More than 250 students and general public attended this talk.

Most of the topics ranging from detector instruments, data analysis, to astrophysics and source modeling were covered in 10 invited talks, 67 contributed oral presentations, and 39 posters.

After the conference, we invited all participants to submit written papers for proceedings. We received 35 papers and went through the peer review process and accepted 34 papers for publications. We are grateful to the referees for their careful review and useful comments.

The 11th Amaldi Conference will be remembered as the last one before the first detection of gravitational wave event, GW150914. We hope that gravitational wave science will become much more mature with this detection and the forthcoming Amaldi conferences will continue to provide a most extensive venue for communication among the scientists working on various aspects of gravitational waves.

Hyung Mok Lee
John Oh
Editors

Journal of Physics: Conference Series **716** (2016) 011002 doi:10.1088/1742-6596/716/1/011002

Peer review statement

All papers published in this volume of *Journal of Physics: Conference Series* have been peer reviewed through processes administered by the proceedings Editors. Reviews were conducted by expert referees to the professional and scientific standards expected of a proceedings journal published by IOP Publishing.

Terrestrial detector for low-frequency gravitational waves based on full tensor measurement

H J Paik[1], M V Moody[1], C E Griggs[1], H M Lee[2] and E Majorana[3]

[1] Department of Physics, University of Maryland, College Park, MD 20742, USA
[2] Department of Physics and Astronomy, Seoul National University, Seoul 151-747, Korea
[3] INFN Sezione di Roma and Università "Sapienza", I-00185 Roma, Italy

Email: hpaik@umd.edu

Abstract. Two serious obstacles in constructing terrestrial gravitational wave (GW) detectors that can resolve low-frequency signals ($\leq$ 10 Hz) are seismic and Newtonian noises. Here we describe a new detector concept by adopting new measurement techniques and configurations to overcome the present low-frequency barrier due to these noises. Six magnetically levitated superconducting test masses, widely separated along three orthogonal axes, each with three degrees of freedom, constitute a tensor GW detector. The tensor outputs could be combined to better reject the Newtonian noise. Unlike current two-dimensional detectors, a single tensor detector is able to determine the polarization of GWs and the direction to sources on its own.

1. Introduction

Several astrophysical processes generate GWs below 10 Hz, and will not be observed by the advanced terrestrial laser interferometer detectors. Two serious obstacles in constructing terrestrial GW detectors that can resolve low-frequency signals are seismic and Newtonian noises (NN). Recently, the possibility of constructing three different types of low-frequency terrestrial GW detector has been discussed [1]. In this paper, we present an alternative design.

Due to the transverse nature of GWs, a detector that measures all the components of the curvature tensor could distinguish GWs from near-field Newtonian gravity. By combining six magnetically levitated superconducting test masses, one could construct a *full-tensor* detector. We discuss a design concept that could reach sensitivity $\leq 10^{-20}$ Hz$^{-1/2}$ at 0.1-10 Hz and discuss the procedure of mitigating the seismic and Newtonian noise. We call this detector SOGRO (Superconducting Omni-directional Gravitational Radiation Observatory).

Because of the finite sizes of white dwarfs (WDs), WD-WD binary merger takes place at $f < 1$ Hz. SOGRO will be able to detect any WD binaries within 1 Mpc. Within this horizon, there are two galaxies: Milky Way Galaxy and Andromeda (M31). The combined merger rate of WD-WD binaries within these two galaxies is estimated to be ~3 per 100 years [1]. If the signal is integrated over one year, the probability of finding a WD-WD binary during 1-year operation of SOGRO is ~30% since each event is expected to last for ~10 years. Direct observation of WD-WD binaries via GW will provide valuable information on the evolution of binary stars composed of relatively low mass stars.

The existence of the intermediate mass black hole (IMBH) is still under vigorous debate, but there are several candidates of IMBH of up to a few tens of thousands of solar mass [2]. Binary mergers composed of IMBHs can be detected by SOGRO up to a few Gpc.

2. Full-tensor GW detector

According to general relativity, a gravitational field is characterized by a curvature tensor. A full-tensor detector could be constructed by measuring five degenerate quadrupole modes of a solid sphere [3]. A tensor detector is sensitive to GWs coming from *any direction* with *any polarization* and is thus capable of resolving the source direction and polarization.

One could construct a low-frequency (0.1 to 10 Hz) tensor GW detector by combining six almost free test masses. Figure 1 shows the test mass configuration of SOGRO. Six superconducting test masses, each with three linear degrees of freedom, are levitated over three orthogonal mounting tubes. The test masses are made of niobium (Nb) in the shape of a rectangular shell with circular flanges. Superconducting levitation/alignment coils and sensing capacitors (not shown) are located in the gap between the test masses and the mounting tubes, as well as on the outer surfaces of the test masses.

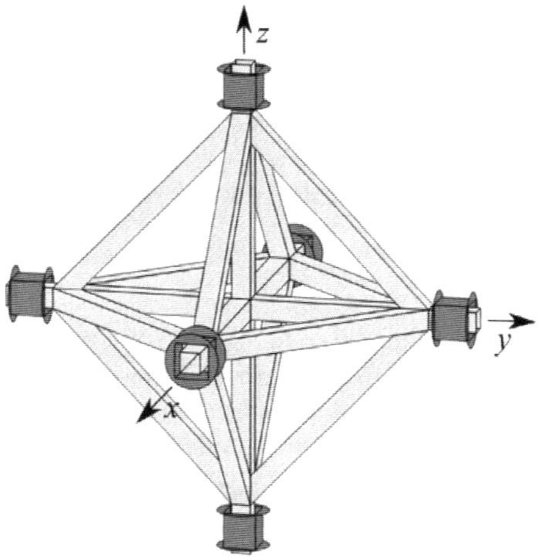

Figure 1. Test mass configuration for SOGRO. Motions of six magnetically levitated test masses are combined to measure all six components of the curvature tensor.

The along-axis motions of the two test masses on each coordinate axis are differenced to measure a diagonal component of the wave. The cross-axis (rotational) motions of the four test masses on each coordinate plane are differenced to measure an off-diagonal component of the wave. The detector also measures the three linear and three angular acceleration signals by *summing* the along-axis and cross-axis test mass motions. These common-mode (CM) signals are used to remove the residual sensitivity of the differential-mode (DM) channels of the detector to the platform accelerations [4].

Since test mass motion is measured with respect to the sensing circuit elements mounted on the platform, SOGRO requires a *rigid platform* with mode frequencies above the signal bandwidth, > 10 Hz. To alleviate excessive demand on cryogenics, the platform must not be too heavy yet sufficiently rigid. Our preliminary FEM analysis shows that, by building the 3D cross with aluminum (Al) square tubes triangulated by Al square tube struts (figure 1), a 30-m SOGRO platform can be constructed with all the mode frequencies > 19 Hz, except for one CM at 15 Hz, and a total mass of ~70 tons.

Magnetic levitation allows the test masses to have horizontal resonance frequencies $f_h \leq 0.01$ Hz. In the vertical direction, the stiffness due to the magnetic field can be made to cancel for the DM by connecting the levitation coils in series such that the total inductance remains constant [4]. In reality, the vertical DM frequencies f_v become ~1 Hz due to nonlinearities in the levitation coils. A "superconducting negative spring" has been demonstrated to reduce the resonance frequency of the SGG [4]. A similar technique could be applied to SOGRO to reduce f_v to 0.01 Hz.

An extremely low-noise transducer with high energy coupling is required for a sensitive GW detector. For SOGRO, we propose to employ a *superconducting tuned capacitor bridge transducer* [5]. Figure 2 shows a transducer for a diagonal-component channel, where $C_{ij}(t)$ is a Nb capacitor plate located near the j-th face of the i-th test mass. The bridge output is coupled to a nearly quantum-limited dc SQUID through a superconducting transformer. The sensing capacitances are modulated by GWs at ω and the bridge is driven at the resonance frequency of the capacitor bridge coupled to the SQUID ω_p, well above the 1/f noise corner frequency of the SQUID. The electrical resonance increas-

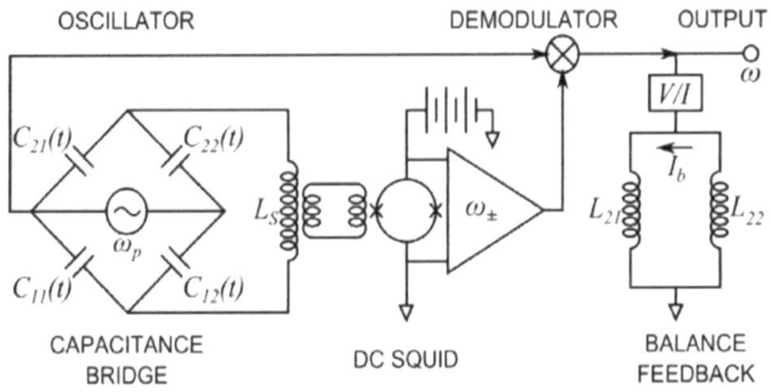

Figure 2. Superconducting tuned capacitor bridge transducer.

es energy coupling constant β. The carrier signal at ω_p is precisely balanced by applying feedback forces to the test masses, eliminating the oscillator noise at the bridge output.

3. Detector sensitivity

The noise power spectral density of a GW detector with an active transducer is given [6] by

$$S_h(f) = \frac{16}{ML^2\omega^4}\left\{\frac{k_B T\omega_D}{Q_D} + \frac{\left|\omega^2-\omega_D^2\right|}{2\omega_p}\left(1+\frac{1}{\beta^2}\right)^{1/2} k_B T_N\right\}, \tag{1}$$

where M and L are the mass of each test mass and the arm-length of the detector; T is the temperature; $\omega_D = 2\pi f_D$ and Q_D are the DM frequency and Q; and T_N is the noise temperature of the SQUID.

Table 1 shows the proposed parameter values and detector noises for two detector options: SOGRO with L = 30 m cooled to T = 1.5 K by pumping on liquid helium or using cryocoolers and advanced SOGRO (aSOGRO) with L = 100 m cooled to 0.1 K by using He^3-He^4 dilution refrigerators. We set $20h$ and $2h$ as the SQUID noise goals for SOGRO and aSOGRO, respectively.

In figure 3, we plot the expected strain sensitivities of SOGRO and aSOGRO, as well as those of advanced LIGO (aLIGO) [7], advanced Virgo (AdV) [8] and eLISA [9]. SOGRO fills the sensitivity gap between advanced terrestrial interferometers and proposed space interferometer. The NN due the Rayleigh and infrasound waves *before mitigation* are also plotted [1]. SOGRO has a detector noise $\leq 10^{-20}$ Hz$^{-1/2}$ at 1-10 Hz. To reach the detector noise limit, the NN must be mitigated by up to 70 dB for SOGRO. The NN mitigation will be discussed in a separate paper in the same proceedings.

4. Seismic noise rejection

We propose to construct SOGRO underground at a depth ≥ 1 km. The seismic noise level of an underground lab at 0.1-10 Hz is $\leq 3\times10^{-7}$ m s^{-2} Hz$^{-1/2}$ [10]. This is 11 orders of magnitude above the target sensitivity of SOGRO. Vibration isolation at such low frequencies is very challenging. The

Table 1. Detector parameters and expected sensitivities of SOGRO and aSOGRO.

Parameter	SOGRO	aSOGRO	Method employed (SOGRO/aSOGRO)
Each test mass M	5 ton	10 ton	Nb square tube
Arm-length L	30 m	100 m	Over "rigid" platform
Antenna temperature T	1.5 K	0.1 K	Superfluid He/dilution refrigerator
Platform temperature T_p	1.5 K	1.5 K	$Q_p = 5\times10^6/10^7$
DM frequency f_D	0.01 Hz	0.01 Hz	Magnetic levitation
DM quality factor Q_D	5×10^8	10^9	Surface polished pure Nb
Signal frequency f	0.1-10 Hz	0.1-10 Hz	
Pump frequency f_p	50 kHz	50 kHz	Tuned capacitor bridge transducer
Amplifier noise number n	20	2	Nearly quantum-limited dc SQUID
Detector noise $S_h^{1/2}(f)$	2×10^{-20} Hz$^{-1/2}$	2×10^{-21} Hz$^{-1/2}$	Computed at 1 Hz

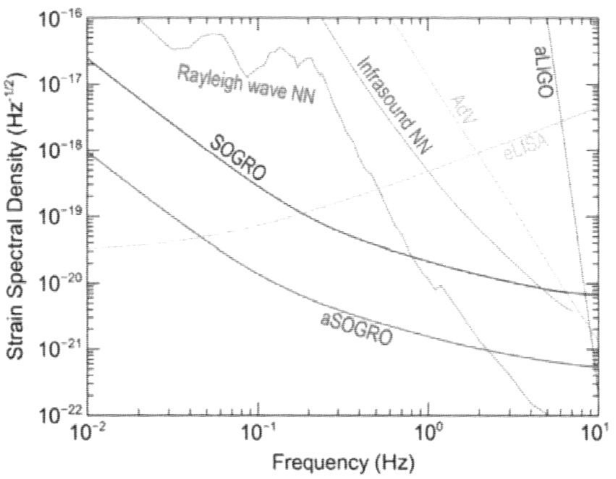

Figure 3. Expected strain sensitivities of SOGRO and aSOGRO in comparison with those of aLIGO, AdV and eLISA.

CM rejection techniques demonstrated with SGGs [11] can be applied to SOGRO and improved to one part in 10^{10}. By precisely matching the accelerometer scale factors and further compensating residual errors, the sensitivity to CM acceleration is reduced to 10^{-9}, lightening the vibration isolation requirement to 20 dB.

The platform is suspended as a long (≥ 15 m) pendulum from its center. The platform will then have horizontal frequencies of ≤ 0.13 Hz and angular frequencies ~1 mHz. This pendulum suspension will provide a passive isolation of 35 dB to horizontal accelerations and 120 dB to angular accelerations at 1 Hz, and much more at 10 Hz. For the vertical direction, 20-dB isolation must be provided by combining passive and active isolation.

5. Conclusions

A wideband *tensor* GW detector with sensitivity $\leq 10^{-20}$ $\text{Hz}^{-1/2}$ for the band 0.1-10 Hz could be constructed by using six widely separated, levitated superconducting test masses. Such a detector would be capable of determining the source direction and wave polarization. The tensor outputs could be combined to better reject the NN.

With SOGRO, one could set a new limit for GW flux at f = 0.1-10 Hz, not covered by the present terrestrial detectors. Binaries composed of two IMBHs of 10^4 solar masses at the distance of a few billion light years or binaries composed of two WDs within the local group can be detected.

Major technical challenges to SOGRO are the construction of a large (30~100 m), rigid, platform that can be cooled to 0.1~1.5 K and obtaining the required high Q (~10^9) in the levitated superconducting test masses. In spite of these challenges, SOGRO deserves serious consideration for a low-frequency terrestrial GW detector since it is a full-tensor detector with unique capabilities.

[1] Harms J *et al* 2013 *Phys. Rev. D* **88** 122003
[2] Gair J R *et al* 2011 *Gen. Rel. Grav.* **43** 485
[3] Wagoner R V and Paik H J 1976 *Proc. Accademia Nazionale dei Lincei Int. Symp. on Experimental Gravitation* (Pavia, Italy) p 257
[4] Moody M V, Chan H A and Paik H J 1986 *J. Appl. Phys.* **60** 4308
[5] Cinquegrana C *et al* 1993 *Phys. Rev. D* **48** 448
[6] Paik H J 1979 *Proc. 2nd Marcel Grossmann Meeting on Recent Developments of General Relativity* (Trieste, Italy)
[7] LIGO Scientific Collaboration 2009 *Rep. Prog. Phys.* **72** 076901
[8] Accadia T *et al* 2011 *Class, Quantum Grav.* **28** 114002
[9] LISA Science Team 2012 *eLISA homepage* http://www.elisa-ngo.org
[10] Beker M G *et al* 2012 *J. Phys. Conf. Ser.* **363** 012004
[11] Moody M V, Paik H J and Canavan E R 2002 *Rev. Sci. Instrum.* **73** 3957

Journal of Physics: Conference Series **716** (2016) 012002 doi:10.1088/1742-6596/716/1/012002

On the distribution of stellar-sized black hole spins

Alex B. Nielsen

Max Planck Institut für Gravitationsphysik, Callinstrasse 38, D-30167 Hannover, Germany and Leibniz Universität Hannover, Welfengarten 1-A, D-30167 Hannover, Germany

Abstract. Black hole spin will have a large impact on searches for gravitational waves with advanced detectors. While only a few stellar mass black hole spins have been measured using X-ray techniques, gravitational wave detectors have the capacity to greatly increase the statistics of black hole spin measurements. We show what we might learn from these measurements and how the black hole spin values are influenced by their formation channels.

1. Introduction

Gravitational wave observations will be sensitive to the spins of black holes [ ]. Simulated search pipelines indicate that accounting for black hole spin can substantially improve the sensitivity of gravitational wave searches for binaries containing one [ ] or two [ ] black holes. However, this improvement comes at a cost of increased computational resources and is only realised if the spins of black holes are indeed significant. It is therefore necessary to balance the increased cost with the expectation that black holes will indeed have large spins. Here we discuss that evidence.

Truly isolated black holes are expected to satisfy the Kerr bound, $a_* \equiv cJ/GM^2 < 1$, where J and M denote the angular momentum and mass of the black hole and c and G the speed of light and Newton's gravitational constant respectively. This bound is manifested in the formula for the spatial coordinate of the event horizon in the Kerr solution, $r = M + M\sqrt{1 - a_*^2}$.

The value of a_* can formally be computed for any object that has a mass and angular momentum and it is worth recalling that the Kerr bound is only a bound on black holes. Many non-compact objects such as the Earth, the Sun and extremely rapidly rotating massive stars like VFTS 102[ ] do not satisfy this bound, whilst compact objects like the rapidly rotating millisecond pulsar PSR J1748-2446ad [ ] and the near-extremal black hole candidate Cygnus X-1[ ] do. This is described in Table 1 where the spin values for the solid objects assume they are constant density spheres.

While the exterior spacetimes of these objects are all approximately vacuum, axisymmetric and stationary as required by the Kerr solution, except for Cygnus X-1 they do not describe black holes (do not contain horizons) and so can have a different multipole structure to the Kerr solution and are not constrained by the Kerr bound.

Things are even more extreme for elementary particles. The quantum spin of an elementary particle can be related to an asymptotic classical angular momentum in the sense of the Einstein-de Haas effect. In this way we can simply calculate values for the mass, specific angular momentum and charge of elementary particles. An electron's mass is 2.3×10^{-66}secs and its specific angular momentum is 6.4×10^{-22} so its a_* value is a whopping 2.8×10^{44}. (In fact things are even more extreme in the context of the Kerr-Newman solution as the electron has a

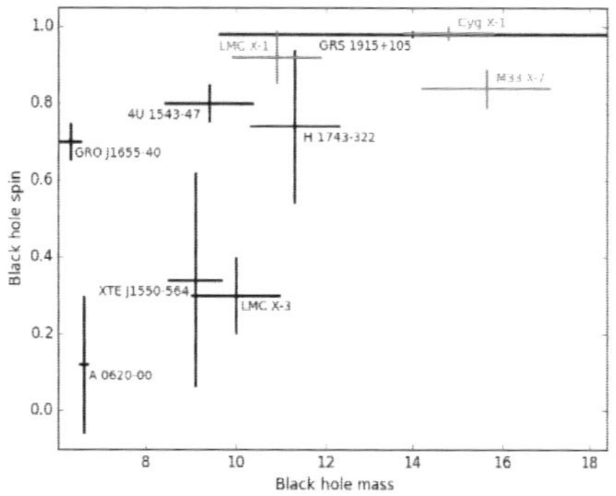

Figure 1. Masses and spins for 10 black holes with approximate error bars. The three high-mass-X-ray-binary systems, LMC X-1, Cygnus X-1 and M33 X-7 are indicated by names above the line and in red online.

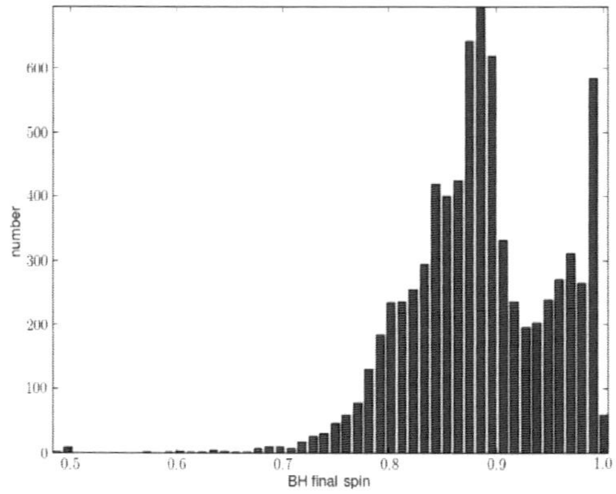

Figure 2. Distribution (upper bound) of spins for black holes in NSBH binaries, with masses greater than $2M_\odot$, assuming NS natal spins of 0, BH natal spins of 0.5 and merging in 15 billion years. Adapted from [ ].

square root charge of 4.6×10^{-45} seconds). The reason for the violation of the Kerr bound in these cases is that elementary particles typically have Planckian values of spin, but not of mass. Of the Standard Model elementary particles, only the Higg's boson satisfies the Kerr bound because its charge and spin are thought to be zero.

Object	Mass [s]	J/M [s]	a_*
Earth	1.5×10^{-11}	1.3×10^{-8}	895
Sun	4.9×10^{-6}	6.1×10^{-6}	1.2
VFTS 102	1.2×10^{-4}	9.3×10^{-3}	75
PSR J1748-2446ad	6.9×10^{-6}	2.9×10^{-6}	0.4
Cygnus X-1	7.30×10^{-5}	7.23×10^{-5}	0.99

Table 1. Approximate values of mass and specific angular momentum for the Earth, Sun, a rapidly spinning massive star VFTS 102, a rapidly spinning neutron star PSR J1748-2446ad and a rapidly spinning black hole Cygnus X-1. For ease of comparison, both the mass and specific angular momentum values are given in seconds.

2. X-ray observations

X-ray observations of accretion disks have been able to measure the spins of around 10 stellar mass black holes [ ]. These are displayed in Fig 1.

Of these most are Low Mass X-ray Binary (LMXB) sources and are unlikely to form the double compact object systems necessary to be seen by the current generation of ground-based gravitational wave detectors. Three of the systems (LMC X-3, M33 X-7 and Cygnus X-1) are High Mass X-ray Binaries (HMXB) where the companion to the black hole is a massive star, with mass greater than $\sim 10M_\odot$ and these systems do have a chance to form either neutron star-black hole binaries or binary black holes systems. These systems in themselves are not targets for current ground-based GW observatories as they are many millions of years away from merging but it is interesting to note that all three of these HMXBs have black holes with large spins $a* > 0.85$ and this suggests that the population of black holes in compact binary systems might be dominated by black holes with large spins. The probability of obtaining three values all above 0.7 from a flat distribution is only 3% .

On the other hand, observations of neutron star pulsars suggest that the spins of neutron stars in compact binaries are very low $a* < 0.05$. If this is indeed the case it implies that the formation channels of these objects are very different in terms of their angular momentum evolution. Neutron stars may be born with low spins or they may spin down rapidly after birth. The observation of young slowly spinning pulsars, such as the Crab pulsar rotating at only ~ 30Hz with a spin down rate of ~ 0.01Hz per year, 1000 years after its birth, constrains models of rapid spin-down for neutron stars.

3. Before collapse - tidal locking

To form a compact binary system requires two large heavy stars in a binary. Single massive stars are typically slowly rotating at the end of their lives dues to stellar wind losses and not differentially rotating due to angular momentum redistribution by magnetic torques [ ]. But stars in binaries can spin up, increasing their rotational angular momentum by mass transfer [ ]. The spin rate of both stars can be affected by tides and mass transfer [ ]. Tidal interactions will tend to circularize the orbits of massive stars[ ]. Massive stars in binaries can have rotational velocities ~ 200km/s [ ].

Close binary systems are expected to be tidally locked [ ]. The tidal locking timescale is between 100 and 10,000 years for helium core stars before they turn into black holes [ ]. The tidal locking is more efficient on the larger object the closer to equal mass the bodies, perhaps giving an explanation why BHs in LMXBs have a wide range of spins, but predominantly high spins in HMXBs. It has been shown that there is a correlation between the black hole mass and orbital period in X-ray transients that may be partly explained by tidal locking [ ]. For simplicity here we will assume that the moment of inertia for each star is given by kmr^2 where k is a constant (equal to 2/5 for solid balls of uniform density). The relationship between radius and mass is often assumed to be $r \sim m^\alpha$ where typical values of α for massive stars are 0.5 to 0.8. Then we find that the dimensionless spin parameter a_* for an object with mass m_1 is given by

$$a_* = k\sqrt{\frac{c^2 r_\odot}{G m_\odot}}\left(\frac{r_\odot}{a}\right)^{3/2}\left(\frac{m_1}{m_\odot}\right)^{2\alpha-1}\sqrt{\frac{m_1}{m_\odot}+\frac{m_2}{m_\odot}}. \quad (1)$$

where a denotes the semi-major axis of the orbit. Note that the three HMXB systems would have $a_* > 1$ if they were tidally locked stars at their current mass and separation values.

Observed rotation rates of massive stars in binaries tend to be significantly higher than the tidal locking rate [ ], since a $10-10$ binary would need a separation of $10r_\odot$ to be tidally locked if the rotational velocity is ~ 300km/s, corresponding to $a_* \sim 100$. Stars that are further apart will have a lower rotational velocity. A separation of 1000 solar radii would have a tidally locked velocity of ~ 300m/s, since $a_* \sim a^{-3/2}$.

4. During collapse - core collapse and supernovae

Different scenarios have been proposed for how collapse proceeds for massive stars. These distinguish between failed core collapse (collapsar) with no explosion and successful core collapse (supernova) where the outer envelope is partially ejected [ ]. Direct collapse may be favoured for heavy systems as the energy flux from the core is not enough to expel the outer shell. If a star collapses directly to a black hole, without shedding any matter and conserving its mass and its angular momentum, then its a_* value would be conserved.

For failed core collapse models, O'Connor and Ott [ ] find a linear relationship between initial specific angular momentum of the star and final black hole spin parameter (their fig 11). The relation is approximately $j = 2.75a_*$. The specific angular momentum before collapse, j,

Journal of Physics: Conference Series **716** (2016) 012002 doi:10.1088/1742-6596/716/1/012002

can be calculated using the tidal lock model above, to obtain a final a_* after collapse of,

$$a_* = 155k \left(\frac{r_\odot}{a}\right)^{3/2} \left(\frac{m}{m_\odot}\right)^{2\alpha+0.5} . \tag{2}$$

It should be noted that for rapidly spinning black holes, centrifugal support will tend to prevent the black hole from accreting large amounts of matter during a collapse scenario. If an initial, near extremal black hole forms, subsequent shells with $a_* > 1$ will not be accreted but expelled [ ]. In this case, as in the successful core collapse, detailed modeling is required to derive final mass and spin values for the remnant.

5. After collapse - common envelope phase

The distribution of upper limits on the final BH spins, assuming natal spins of zero for neutron stars and 0.5 for black holes, for merging NSBH systems in [ ] is shown in Fig. 2. This shows that if accretion in the common envelope phase does behave as a thin disk, then the final spins of BHs in NSBH binaries may be quite large, with most values falling in the range 0.8 to 1.0.

The distribution in Fig. 2 actually contains two distinct populations with different formation channels. Most of the high spin systems are actually light black holes that formed initially after collapse as neutron stars and subsequently accreted sufficient material during common envelope to form black holes, where the threshold for black hole formation is set at $2M_\odot$. For these light systems, the accretion of a small amount of matter is enough to spin them up considerably.

The other distribution consists of objects that collapsed to black holes before the common envelope phase. These systems are typically more massive and hence do not spin up as much during common envelope.

6. Conclusion

The limited X-ray evidence suggests that black holes in compact binaries may have high spins. We have discussed several mechanisms that could lead to these high spins, including tidal locking of massive progenitor stars, rapid collapse to black holes and common envelope accretion. There are still large uncertainties of all these processes, although they tend to reinforce one another. Measuring the spins of a large number of black holes with gravitational wave detectors will help to constrain some of these model uncertainties.

Acknowledgments

This work was supported by the Max Planck Society. The author would like to thank Richard O'Shaughnessy for helpful conversations and sharing data for Fig. 2.

References

[1] E. Abdikamalov, S. Gossan, A. M. DeMaio and C. D. Ott, arXiv:1311.3678 [astro-ph.SR].
[2] T. Dal Canton *et al.*, Phys. Rev. D **90** (2014) 8, 082004 [arXiv:1405.6731 [gr-qc]].
[3] C. Cutler *et al.*, Phys. Rev. Lett. **70** (1993) 2984 [astro-ph/9208005].
[4] P. L. Dufton *et al.*, Astrophys. J. **743** (2011) L22 [arXiv:1111.0157 [astro-ph.SR]].
[5] L. Gou, J. E. McClintock, M. J. Reid, J. A. Orosz, J. F. Steiner, R. Narayan, J. Xiang and R. A. Remillard *et al.*, Astrophys. J. **742** (2011) 85 [arXiv:1106.3690 [astro-ph.HE]].
[6] J. W. T. Hessels, S. M. Ransom, I. H. Stairs, P. C. C. Freire, V. M. Kaspi and F. Camilo, Science **311** (2006) 1901 [astro-ph/0601337].
[7] J. R. Hurley, C. A. Tout and O. R. Pols, Mon. Not. Roy. Astron. Soc. **329** (2002) 897 [astro-ph/0201220].
[8] K. R. Lang, *Essential Astrophysics*, Springer-Verlag, Berlin (2013)
[9] C. H. Lee, G. E. Brown and R. A. M. J. Wijers, Astrophys. J. **575** (2002) 996 doi:10.1086/341349 [astro-ph/0109538].
[10] S. E. de Mink, N. Langer, R. G. Izzard, H. Sana and A. de Koter, Astrophys. J. **764** (2013) 166 [arXiv:1211.3742 [astro-ph.SR]].

Journal of Physics: Conference Series **716** (2016) 012002 doi:10.1088/1742-6596/716/1/012002

[11] J. E. McClintock *et al.*, Class. Quant. Grav. **28** (2011) 114009 [arXiv:1101.0811 [astro-ph.HE]].
[12] E. M. Mendez, MNRAS **413**, (2011) 183 arXiv:1011.6385 [astro-ph.HE].
[13] E. O'Connor and C. D. Ott, Astrophys. J. **730** (2011) 70 [arXiv:1010.5550 [astro-ph.HE]].
[14] R. W. O'Shaughnessy, J. Kaplan, V. Kalogera and K. Belczynski, Astrophys. J. **632** (2005) 1035 [astro-ph/0503219].
[15] S. Privitera *et al.*, Phys. Rev. D **89** (2014) 2, 024003 [arXiv:1310.5633 [gr-qc]].
[16] H. Sana, A. de Koter, S. E. de Mink, P. R. Dunstall, C. J. Evans, V. Henault-Brunet, J. M. Apellaniz and O. H. Ramirez-Agudelo *et al.*, arXiv:1209.4638 [astro-ph.SR].
[17] S. E. Woosley, Astrophys. J. **405** (1993) 273.

Distinguishing transient signals and instrumental disturbances in semi-coherent searches for continuous gravitational waves with line-robust statistics

David Keitel

Albert-Einstein-Institut, Callinstr. 38, 30167 Hannover, Germany

E-mail: `david.keitel@ligo.org`

Abstract. Non-axisymmetries in rotating neutron stars emit quasi-monochromatic gravitational waves. These long-duration 'continuous wave' signals are among the main search targets of ground-based interferometric detectors. However, standard detection methods are susceptible to false alarms from instrumental artefacts that resemble a continuous-wave signal. Past work [Keitel, Prix, Papa, Leaci and Siddiqi 2014, *Phys. Rev. D* **89** 064023] showed that a Bayesian approach, based on an explicit model of persistent single-detector disturbances, improves robustness against such artefacts. Since many strong outliers in semi-coherent searches of LIGO data are caused by transient disturbances that last only a few hours or days, I describe in a recent paper [Keitel D 2015, LIGO-P1500159] how to extend this approach to cover transient disturbances, and demonstrate increased sensitivity in realistic simulated data. Additionally, neutron stars could emit transient signals which, for a limited time, also follow the continuous-wave signal model. As a pragmatic alternative to specialized transient searches, I demonstrate how to make standard semi-coherent continuous-wave searches more sensitive to transient signals. Focusing on the time-scale of a single segment in the semi-coherent search, Bayesian model selection yields a simple detection statistic without a significant increase in computational cost. This proceedings contribution gives a brief overview of both works.

1. Introduction

Rotating neutron stars (NSs) with non-axisymmetric deformations emit long-term stable quasi-monochromatic gravitational radiation, called *continuous waves* (CWs). [1] These are searched for with data from terrestrial interferometers, such as LIGO and Virgo. A standard detection method is the matched-filter $\mathcal{F}$*-statistic* [2, 3]. Large-parameter space blind searches are most efficient with *semi-coherent* methods [4, 5, 6], splitting the data into shorter segments.

CW detection methods, such as the semi-coherent $\mathcal{F}$-statistic, are vulnerable to deviations from the standard Gaussian noise model that more closely resemble the quasi-monochromatic signal model. The data typically contains many narrow spectral artefacts, called *lines*, of instrumental or environmental origin. [7, 8, 9] These produce strong outliers in the $\mathcal{F}$-statistic, thus requiring some form of vetoing or follow-up procedure. [10, 11, 12, 13]

In previous work [14, 15], it was shown how a Bayesian model-comparison approach [16, 17] can be used to derive a *line-robust statistic* that achieves the same sensitivity as the $\mathcal{F}$-statistic in quiet, almost-Gaussian noise, but is significantly less affected by lines.

A recent extension of this approach [18] includes models for *transient* artefacts and transient signals. Transient line-like disturbances with typical durations of a few hours have been found in LIGO data [11, 12, 19] and cause similar problems as persistent lines. In addition, NSs can also emit transient signals that follow the CW model over time-scales of hours to days (*tCWs*, see [17, 20]). So a general detection statistic should be robust to both persistent and transient lines, while retaining or improving sensitivity to persistent and transient signals.

In Sec. 2, I give a brief review of the line-robust statistics of [14] for the case of persistent CW signals and persistent line disturbances, Then I sketch the approach of [18] to addressing transient disturbances and signals in Sec. 3, without any mathematical details. In both cases, I present results of injection studies with either archival LIGO S5 data or simulated data.

2. Line-robust statistics for persistent signals and disturbances

In the Bayesian hypothesis testing framework, the multi-detector $\mathcal{F}$-statistic [2, 3] is derived from the posterior odds between a CW signal hypothesis $\widehat{\mathcal{H}}_{\mathrm{S}} : \boldsymbol{x}(t) = \boldsymbol{n}(t) + \boldsymbol{h}(t;\mathcal{A},\lambda)$ and a pure Gaussian noise hypothesis $\widehat{\mathcal{H}}_{\mathrm{G}} : \boldsymbol{x}(t) = \boldsymbol{n}(t)$. Here, $\boldsymbol{n}(t)$ is a multi-detector Gaussian time series and $\boldsymbol{h}(t;\mathcal{A},\lambda)$ is a CW waveform with amplitude parameters $\mathcal{A}$ and phase-evolution parameters λ. As first found in [16] and extended to the semi-coherent case in [17], the result after marginalisation over $\mathcal{A}$ is $\widehat{O}_{\mathrm{S/G}}(\boldsymbol{x}) \equiv P\left(\widehat{\mathcal{H}}_{\mathrm{S}}\middle|\boldsymbol{x},\mathcal{I}\right) / P\left(\widehat{\mathcal{H}}_{\mathrm{G}}\middle|\boldsymbol{x},\mathcal{I}\right) \propto \mathrm{e}^{\widehat{\mathcal{F}}(\boldsymbol{x})}$.

The simple line model of [14] describes a single-detector line by $\widehat{\mathcal{H}}_{\mathrm{S}}$ restricted to that detector (indexed by X), $\widehat{\mathcal{H}}_{\mathrm{L}}^X : x^X(t) = n^X(t) + h^X(t;\mathcal{A}^X)$. A hypothesis test of 'CW signal vs. either Gaussian noise or a line' yields posterior odds

$$\widehat{O}_{\mathrm{S/GL}}(\boldsymbol{x}) = \frac{P\left(\widehat{\mathcal{H}}_{\mathrm{S}}\middle|\boldsymbol{x},\mathcal{I}\right)}{P\left(\widehat{\mathcal{H}}_{\mathrm{G}}\middle|\boldsymbol{x},\mathcal{I}\right) + P\left(\widehat{\mathcal{H}}_{\mathrm{L}}\middle|\boldsymbol{x},\mathcal{I}\right)} \,. \tag{1}$$

Introducing the simplest example case of 2 detectors $X = 1,2$ and 2 segments $\ell = 1,2$, as well as a pictorial presentation of $\square$ for pure Gaussian noise in the data sub-set (X,ℓ) and $\blacksquare$ for a signal or disturbance, I can also represent the standard $\mathcal{F}$-statistic and (1) as

$$\mathrm{e}^{\widehat{\mathcal{F}}} \propto \frac{P(\begin{smallmatrix}\blacksquare&\blacksquare\\\blacksquare&\blacksquare\end{smallmatrix} \mid \boldsymbol{x})}{P(\begin{smallmatrix}\square&\square\\\square&\square\end{smallmatrix} \mid \boldsymbol{x})} \quad \text{and} \quad O_{\mathrm{S/GL}}(\boldsymbol{x}) = \frac{P(\begin{smallmatrix}\blacksquare&\blacksquare\\\blacksquare&\blacksquare\end{smallmatrix} \mid \boldsymbol{x})}{P(\begin{smallmatrix}\square&\square\\\square&\square\end{smallmatrix} \mid \boldsymbol{x}) + P(\begin{smallmatrix}\blacksquare&\square\\\blacksquare&\square\end{smallmatrix} \mid \boldsymbol{x}) + P(\begin{smallmatrix}\square&\blacksquare\\\square&\blacksquare\end{smallmatrix} \mid \boldsymbol{x})} \,. \tag{2}$$

See [14] for the derivation, tuning methods for the free parameters and detailed tests. As a detection statistic, one can always replace the posterior odds by a corresponding Bayes factor that factors out the prior odds, e.g. $B_{\mathrm{S/GL}}(\boldsymbol{x},\mathcal{I}) \equiv O_{\mathrm{S/GL}}(\boldsymbol{x},\mathcal{I})/o_{\mathrm{S/GL}}(\mathcal{I})$.

Fig. 1, reproduces some of the main results from an injection study in [14], searching for simulated persistent CW signals in archival LIGO S5 data. These demonstrate that a properly tuned $B_{\mathrm{S/GL}}$ behaves just like $\widehat{\mathcal{F}}$ in quiet data (left panel), but improves over both $\widehat{\mathcal{F}}$ and the established $\mathcal{F}$-*stat consistency veto* [10, 11, 13] in the presence of lines (right panel).

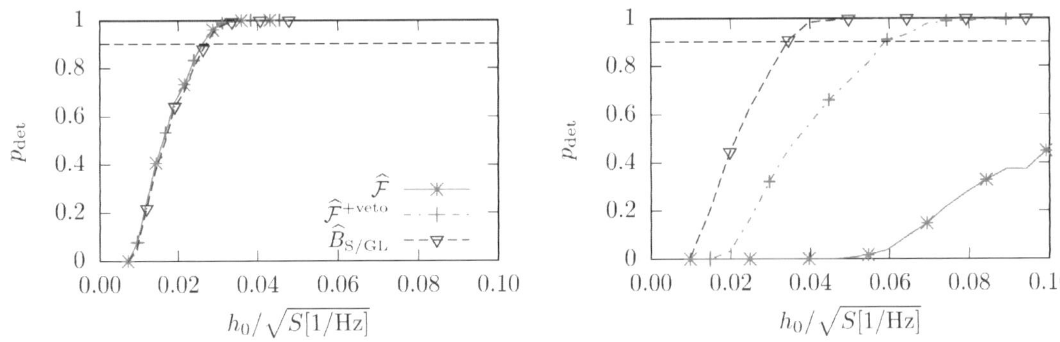

Figure 1. Detection efficiency p_{det} for persistent CW signals in LIGO S5 data, as a function of PSD-scaled signal amplitude $h_0/\sqrt{S}$, for the following semi-coherent statistics: $\widehat{\mathcal{F}}$ [2, 3], $\widehat{\mathcal{F}}^{+\mathrm{veto}}$ with consistency veto [10, 11], $\widehat{B}_{\mathrm{S/GL}}$ from Eq. (1) [14]. Left panel: a quiet 50 mHz band starting at 54.20 Hz, right panel: a disturbed 50 mHz band starting at 69.70 Hz. The dashed horizontal line marks $p_{\mathrm{det}} = 90\%$. Statistical uncertainties are smaller than the plot markers; some markers are omitted to make curves more discernible. Figures reproduced with data from [14].

3. Addressing transient disturbances and signals

To address transient effects, [18] adds a new 'transient-line hypothesis' for a quasi-harmonic disturbance in a single segment ℓ and single detector X: $\widetilde{\mathcal{H}}_{\mathrm{tL}}^{X\ell} : x^{X\ell}(t) = n^{X\ell}(t) + h^{X\ell}(t;\mathcal{A}^{X\ell})$, and a multi-detector-coherent transient-signal hypothesis $\widetilde{\mathcal{H}}_{\mathrm{tS}}^{\ell} : \boldsymbol{x}^{\ell}(t) = \boldsymbol{n}^{\ell}(t) + \boldsymbol{h}^{\ell}(t;\mathcal{A}^{\ell})$.

Testing persistent CWs against all three types of noise (Gaussian, persistent line, transient line) then leads to posterior odds

$$\widehat{O}_{\mathrm{S/GLtL}}(\boldsymbol{x}) = \frac{P\left(\widehat{\mathcal{H}}_{\mathrm{S}}\middle|\boldsymbol{x},\mathcal{I}\right)}{P\left(\widehat{\mathcal{H}}_{\mathrm{G}}\middle|\boldsymbol{x},\mathcal{I}\right) + P\left(\widehat{\mathcal{H}}_{\mathrm{L}}\middle|\boldsymbol{x},\mathcal{I}\right) + P\left(\widehat{\mathcal{H}}_{\mathrm{tL}}\middle|\boldsymbol{x},\mathcal{I}\right)}. \tag{3}$$

or, pictorially in the simplest example case,

$$\widehat{O}_{\mathrm{S/GLtL}}(\boldsymbol{x}) = \frac{P\left(\begin{smallmatrix}\blacksquare&\blacksquare\\\blacksquare&\blacksquare\end{smallmatrix}\mid\boldsymbol{x}\right)}{P\left(\begin{smallmatrix}\square&\square\\\square&\square\end{smallmatrix}\mid\boldsymbol{x}\right) + P\left(\begin{smallmatrix}\blacksquare&\square\\\blacksquare&\square\end{smallmatrix}\mid\boldsymbol{x}\right) + P\left(\begin{smallmatrix}\square&\blacksquare\\\square&\blacksquare\end{smallmatrix}\mid\boldsymbol{x}\right) + P\left(\begin{smallmatrix}\blacksquare&\square\\\square&\square\end{smallmatrix}\mid\boldsymbol{x}\right) + P\left(\begin{smallmatrix}\square&\blacksquare\\\square&\square\end{smallmatrix}\mid\boldsymbol{x}\right) + \ldots}. \tag{4}$$

An even more general test for the possibility of either type of signal (persistent CW or transient tCW) against the same combined noise model yields posterior odds

$$\widehat{O}_{\mathrm{StS/GLtL}}(\boldsymbol{x}) = \frac{P\left(\widehat{\mathcal{H}}_{\mathrm{S}}\middle|\boldsymbol{x},\mathcal{I}\right) + P\left(\widehat{\mathcal{H}}_{\mathrm{tS}}\middle|\boldsymbol{x},\mathcal{I}\right)}{P\left(\widehat{\mathcal{H}}_{\mathrm{G}}\middle|\boldsymbol{x},\mathcal{I}\right) + P\left(\widehat{\mathcal{H}}_{\mathrm{L}}\middle|\boldsymbol{x},\mathcal{I}\right) + P\left(\widehat{\mathcal{H}}_{\mathrm{tL}}\middle|\boldsymbol{x},\mathcal{I}\right)} \tag{5}$$

$$= \frac{P\left(\begin{smallmatrix}\blacksquare&\blacksquare\\\blacksquare&\blacksquare\end{smallmatrix}\mid\boldsymbol{x}\right) + P\left(\begin{smallmatrix}\blacksquare&\blacksquare\\\square&\square\end{smallmatrix}\mid\boldsymbol{x}\right) + P\left(\begin{smallmatrix}\square&\square\\\blacksquare&\blacksquare\end{smallmatrix}\mid\boldsymbol{x}\right)}{P\left(\begin{smallmatrix}\square&\square\\\square&\square\end{smallmatrix}\mid\boldsymbol{x}\right) + P\left(\begin{smallmatrix}\blacksquare&\square\\\blacksquare&\square\end{smallmatrix}\mid\boldsymbol{x}\right) + P\left(\begin{smallmatrix}\square&\blacksquare\\\square&\blacksquare\end{smallmatrix}\mid\boldsymbol{x}\right) + P\left(\begin{smallmatrix}\blacksquare&\square\\\square&\square\end{smallmatrix}\mid\boldsymbol{x}\right) + P\left(\begin{smallmatrix}\square&\blacksquare\\\square&\square\end{smallmatrix}\mid\boldsymbol{x}\right) + P\left(\begin{smallmatrix}\square&\square\\\blacksquare&\square\end{smallmatrix}\mid\boldsymbol{x}\right) + P\left(\begin{smallmatrix}\square&\square\\\square&\blacksquare\end{smallmatrix}\mid\boldsymbol{x}\right)}.$$

The corresponding Bayes factors $\widehat{B}_{\mathrm{S/GLtL}}$ and $\widehat{B}_{\mathrm{StS/GLtL}}$ are again obtained by factoring out the prior odds. For full expressions of these detection statistics, a detailed derivation and a description of the injection studies, the reader is referred to [18].

Fig. 2 shows results of an injection study similar to those in [14] and Fig. 1, using fully simulated data: Gaussian noise, a single-segment disturbance in one detector, and either persistent CW signals (left panel) or tCW signals that last for exactly one segment (right panel).

These tests demonstrate that, with appropriate tuning, the detection statistic $\widehat{B}_{\mathrm{StS/GLtL}}$ is indeed both *robust against transient or persistent disturbances* and *sensitive to transient or*

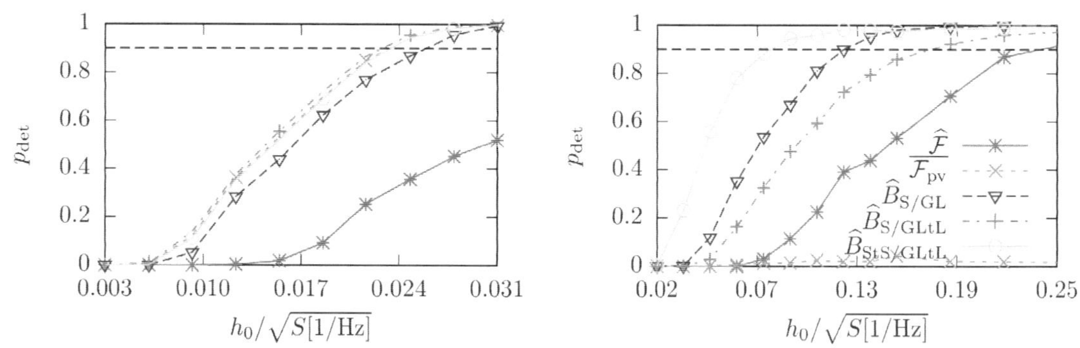

Figure 2. Detection efficiency p_{det} for different signal types in simulated data, for the following statistics: $\widehat{\mathcal{F}}$ [2, 3], $\overline{\mathcal{F}}_{\mathrm{pv}}$ with permanence veto [12], $\widehat{B}_{\mathrm{S/GL}}$ from Eq. (1) [14], $\widehat{B}_{\mathrm{S/GLtL}}$ from Eq. (3) [18], $\widehat{B}_{\mathrm{StS/GLtL}}$ from Eq. (5) [18]. In both panels, the underlying noise is Gaussian with a transient single-segment disturbance. Left panel: persistent CW signals, right panel: transient tCw signals with duration $T_{\mathrm{inj}} = T_{\mathrm{seg}} = 60\mathrm{h}$ in random segments. Plot properties as in Fig. 1.

persistent CW-like signals. In the first aspect, it reproduces the performance of a previously-introduced *permanence veto* [12], while in the latter it avoids the near-total dismissal of tCWs by that veto and improves over existing semi-coherent detection statistics.

As long as transient disturbances or signals are short enough to have a large part of their effect within a single segment, $\widehat{B}_{\mathrm{StS/GLtL}}$ is well approximated by considering only the maximum term in all sums over segments. This means that the search application only needs to store the *loudest* single-segment $\mathcal{F}^{\ell}$, $\mathcal{F}^{X\ell}$. Hence, it is easy to modify existing semi-coherent CW searches, and the extra computational cost and memory requirements will be low. This approach is thus promising for a cheap transient search as an 'add-on' to existing search pipelines, such as Einstein@Home [21]; an option that should be all the more valuable since no dedicated transient-CW search has been performed on real LIGO data so far.

As an end note to this proceeding, let me point out another approach to the detection of transient GWs on similar time-scales, based not on an explicit CW-like model, but on the 'radiometer' search [22]. It should be interesting to compare the sensitivity of both approaches, and of the transient-optimized matched-filter statistic of [17], on realistic data and with various signal characteristics.

Acknowledgments
I thank Reinhard Prix for valuable advice throughout this project; B. Behnke, M.A. Papa, O. Piccini, I. Di Palma and H.-B. Eggenstein for their data-analysis work on LIGO S5 and S6 data that inspired this study; the LIGO Scientific Collaboration for providing those data sets; G. Woan, I.S. Heng and E. Thrane for inspiring discussions about transient signals, and S. Walsh for advice on injection procedures. The injection studies were carried out on the ATLAS cluster at AEI Hannover. This paper has been assigned LIGO document number LIGO-P1500160.

References

[1] Prix R (for the LSC) 2009 Gravitational waves from spinning neutron stars *Neutron Stars and Pulsars* (*Astrophysics and Space Science Library* vol 357) ed Becker W (Springer Berlin Heidelberg) p 651 ISBN 978-3-540-76964-4 (LIGO-P060039-v3) URL `https://dcc.ligo.org/LIGO-P060039/public`
[2] Jaranowski P, Królak A and Schutz B F 1998 *Phys. Rev. D* **58** 063001 (*Preprint* gr-qc/9804014)
[3] Cutler C and Schutz B F 2005 *Phys. Rev. D* **72** 063006 (*Preprint* gr-qc/0504011)
[4] Brady P R and Creighton T 2000 *Phys. Rev. D* **61** 082001 (*Preprint* gr-qc/9812014)
[5] Cutler C, Gholami I and Krishnan B 2005 *Phys. Rev. D* **72** 042004 (*Preprint* gr-qc/0505082)
[6] Prix R and Shaltev M 2012 *Phys. Rev. D* **85** 084010 (*Preprint* 1201.4321)
[7] Aasi J *et al.* (LVC) 2012 *Class. Quant. Grav.* **29** 155002 (*Preprint* 1203.5613)
[8] Accadia T *et al.* 2012 *J. Phys. Conf. Ser.* **363** 012037
[9] Aasi J *et al.* (LVC) 2015 *Class. Quant. Grav.* **32** 115012 (*Preprint* 1410.7764)
[10] Aasi J *et al.* (LVC) 2013 *Phys. Rev. D* **87** 042001 (*Preprint* 1207.7176)
[11] Aasi J *et al.* (LVC) 2013 *Phys. Rev. D* **88** 102002 (*Preprint* 1309.6221)
[12] Behnke B, Papa M A and Prix R 2015 *Phys. Rev. D* **91** 064007 (*Preprint* 1410.5997)
[13] Leaci P 2015 *Phys. Scripta* **90** 125001
[14] Keitel D, Prix R, Papa M A, Leaci P and Siddiqi M 2014 *Phys. Rev. D* **89** 064023 (*Preprint* 1311.5738)
[15] Keitel D and Prix R 2015 *Class. Quant. Grav.* **32** 035004 (*Preprint* 1409.2696)
[16] Prix R and Krishnan B 2009 *Class. Quant. Grav.* **26** 204013 (*Preprint* 0907.2569)
[17] Prix R, Giampanis S and Messenger C 2011 *Phys. Rev. D* **84** 023007 (*Preprint* 1104.1704)
[18] Keitel D 2015 *ArXiv e-prints* Tech. rep. LIGO-P1500159 (*Preprint* 1509.02398)
[19] Piccinni O 2014 *Mitigation of transient disturbances in wide parameter space searches for continuous gravitational wave signals* Master's thesis Università di Roma La Sapienza
[20] Santiago Prieto R I 2014 *Transient Gravitational Waves at r-mode Frequencies from Neutron Stars* Ph.D. thesis University of Glasgow
[21] Allen B *et al.* Einstein@Home distributed computing project URL `http://einsteinathome.org/`
[22] Thrane E, Mandic V and Christensen N 2015 *Phys. Rev. D* **91** 104021 (*Preprint* 1501.06648)

Optical testbed for the LISA phasemeter

T S Schwarze, G Fernández Barranco, D Penkert, O Gerberding, G Heinzel and K Danzmann

Max-Planck-Institut für Gravitationsphysik (Albert-Einstein-Institut), and Institut für Gravitationsphysik, Leibniz Universität Hannover, Callinstrasse 38, 30167 Hannover, Germany

E-mail: thomas.schwarze@aei.mpg.de

Abstract. The planned spaceborne gravitational wave detector LISA will allow the detection of gravitational waves at frequencies between 0.1 mHz and 1 Hz. A breadboard model for the metrology system aka the phasemeter was developed in the scope of an ESA technology development project by a collaboration between the Albert Einstein Institute, the Technical University of Denmark and the Danish industry partner Axcon Aps. It in particular provides the electronic readout of the main interferometer phases besides auxiliary functions. These include clock noise transfer, ADC pilot tone correction, inter-satellite ranging and data transfer. Besides in LISA, the phasemeter can also be applied in future satellite geodesy missions. Here we show the planning and advances in the implementation of an optical testbed for the full metrology chain. It is based on an ultra-stable hexagonal optical bench. This bench allows the generation of three unequal heterodyne beatnotes with a zero phase combination, thus providing the possibility to probe the phase readout for non-linearities in an optical three signal test. Additionally, the utilization of three independent phasemeters will allow the testing of the auxiliary functions. Once working, components can individually be replaced with flight-qualified hardware in this setup.

1. Introduction

In order to detect gravitational waves in the low frequency regime (mHz-Hz), the planned space mission LISA (laser interferometer space antenna) [1, 2] will utilize inter-satellite heterodyne interferometry to detect distance changes between free-floating test masses [3]. The corresponding metrology system needs to provide an interferometer phase readout precision in the order of $\mu\mathrm{rad}/\sqrt{\mathrm{Hz}}$. Hardware to solve this task, a LISA phasemeter, was developed within the scope of an ESA technology development project by a collaboration between the Albert Einstein Institute, the Technical University of Denmark and the Danish industry partner Axcon Aps. It includes the actual heterodyne readout itself together with auxiliary functions to ensure its required high precision and provide additional information for post processing. Their necessity and requirements are dictated by the LISA design and its sensitivity. This will be explained in the following section. In addition to LISA, the phasemeter could also be used in other applications, like geodesy missions [4], or MHz heterodyne interferometry in general.

2. Requirements for the phasemeter

The distance between the LISA satellites will be of the order of several million km. This distance is too long to receive sufficient power from a dispersing laser beam which is directly reflected from one of the remote spacecraft. To solve this problem, a local high-power copy of the beam is created. For this purpose, the received laser beam is mixed with a local laser on each spacecraft. The resulting beatnote is used as a sensor signal to phaselock the local laser to the received beam, thus creating the local high-power copy. The latter is then transmitted back to the opposite satellite. As the LISA constellation will fly on orbits with line-of-sight velocities between satellites in the order of several m/s, Doppler shifts in the order of MHz with both signs shift the beatnote up and down. To avoid zero crossing, a frequency offset must be applied in the lock. These circumstances lead to the requirements for the phasemeter to operate with the earlier stated precision within the MHz regime for the phase readout and to be able to control the local laser. The phasemeter presented in this paper handles the phase readout with a DPLL (digital phase locked loop) [5, 6], which tracks the MHz signal by mixing it with a digital copy. This also allows the implementation of the local laser control in the digital domain.

However, the digital systems and in particular the digital copies of the beatnotes on the satellites, are driven by USOs (ultra stable oscillators), also referred to as clocks in the following. They individually come with a timing jitter and the subsequent differential timing noise between different satellites distorts the phase measurement. This issue is addressed by measuring this differential noise by transferring and comparing the single clock noises, so the differential noise is known and can be removed in post processing [7, 8]. The actual implementation of this concept here is to use a GHz USO, which can directly be imprinted on the main carrier as a sideband and mixed down with its counterpart on a remote satellite [9]. The phasemeter therefore needs be able to downsample the GHz signal into the MHz domain to use it as a digital clock as well as tracking the sideband-sideband beatnote. The former is achieved with dedicated digital hardware called the clock module, while for the latter a DPLL is used. It should be mentioned here, that recent publications also show an alternative option for the clock noise removal [10].

A similar distortion occurs through the non-perfect analog-to-digital conversion. Jitter in its sampling times adds phase noise. However, it can be mitigated by adding a well-known pilot tone to the actual signal as a reference to track the distortion. The pilot tone is derived from the USO and therefore its distribution is incorporated on the clock module.

Finally, the most severe limitation of the LISA sensitivity is introduced by the huge differences in the interferometer arm lengths which lead to amplification of laser frequency noise about several orders of magnitude. To address this issue, the post-processing technique TDI (time delay interferometry) has been developed [8, 11, 12]. It makes use of the fact, that each triangular interferometer formed by the LISA constellation is broken down into measurements between single arms. These single arm measurements can then be artificially delayed during data analysis to compensate for the unequal arm length and to synthesis an equal arm length interferometer. However this requires the knowledge of the absolute arm length and the relative satellite distances, respectively, with sub-meter accuracy. Only then can the correct delays be applied in the TDI algorithm. The phasemeter implements a solution for this inter-satellite ranging [13, 14] by comparing locally generated PRN (pseudo random noise) codes with received ones from the remote spacecraft. The transmission is again achieved by sideband modulation of the main carrier. In a similar manner, data bits can be imprinted on the main carrier for data communication between the spacecraft. A desciption of the actual hardware and details of the implementations can be looked up in [15, 16].

3. Electrical testing

Naturally, all previously mentioned functions of the phasemeter have to be tested. In particular, a sufficient phase readout performance had to be shown for the successful completion of the ESA project, within which frame the phasemeter was developed. This was done by a null measurement with LISA-like signals, which were generated with a dedicated digital signal simulator (DSS). Performing a null measurement means splitting a given signal and feeding it to separate phasemeter channels. The difference in the measured phases then yields the estimate of the readout performance and can be utilized to search for noise sources in the readout channels. Null measurements carried out utilizing the pilot tone correction described in the last section showed sufficient performance below the requirement of $8.5\,\mu\mathrm{rad}/\sqrt{\mathrm{Hz}}$ for frequencies up to 25 MHz [16].

4. Advanced optical testbed - a hexagonal interferometer

Despite the proof of performance with the described electrical null measurement, a more advanced testbed is currently under construction at the Albert Einstein Institute. The leading thought for it is to be able to perform a three signal measurement utilizing optical signals [5]. In contrast to a null measurement, the three signal measurement provides the possibility to search for non-linearities common in the phasemeter channels. This is achieved by generating three signals that, while being different from each other, cancel out when adding their phases. Performing this scheme with optical signals puts it closer to a realistic environment. Besides the test of the phase readout itself, the testbed also provides the capability to test the whole LISA metrology chain including the auxiliary functions.

In detail, the testbed consists of a laser preparation, a hexagonal ultra-stable optical bench as well as the unit under test, the phasemeter itself. At the heart of the testbed is the hexagonal optical bench, a sketch of which can be seen in Fig. 1. It consists of a Zerodur® baseplate and six beamsplitters and beamcombiners, respectively. They were positioned in a way that allows the pairwise mixing of three beams under ultra-stable conditions, effectively implementing an optical three signal test. The beatnotes generated are read out by external photodetectors and fed via cables to the phasemeter. To mitigate angular jitter from the fiber injectors, special quasi-monolithic FIOSs (fiber injector optical assemblies) are under construction and about to be bonded to the bench. Furthermore, the bench will be deployed in a vacuum tank with thermal shielding to avoid distortions due to air fluctuations and to reduce temperature changes.

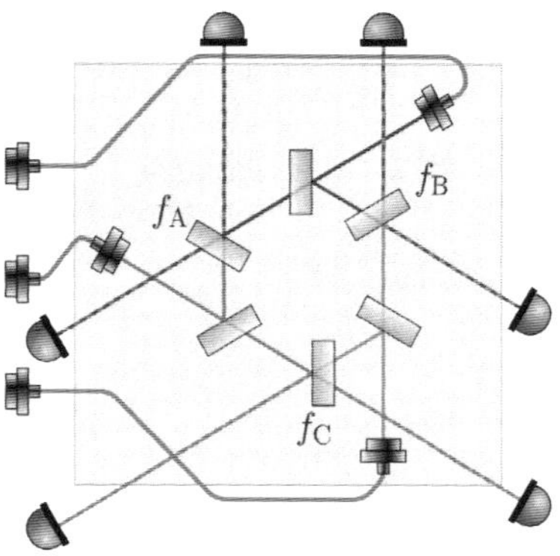

Figure 1. Sketch of the hexagonal ultra-stable optical bench. Three beams injected via fiber injectors are mixed pairwise to implement an optical three signal test.

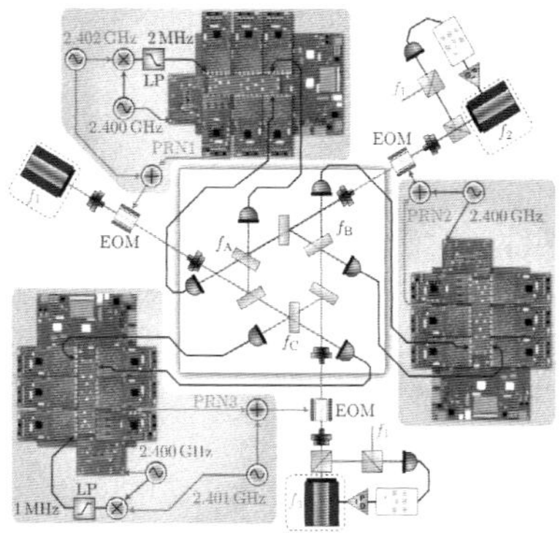

Figure 2. Sketch of the planned final setup, showing three phasemeters and three EOMs with exemplary frequencies for tests of the auxillary functions.

To make sure that the beatnotes generated on the ultra-stable optical bench stay within the bandwidth of the phasemeter, the associated lasers have to be controlled. The final goal is to do this with the same LISA phasemeter used for the actual read out. As an intermediate step, a laser preparation bench has been built, on which once more beatnotes between the lasers are generated and used for locking. This provides the possibility to control and lock the lasers independent from the readout. The loop gain of the laser locks can intentionally be changed to vary the laser frequency noise to show phasemeter performance under different conditions.

To be able to also test the auxiliary function of the phasemeter, it is planned to use three individual phasemeters together with a dedicated GHz source and an EOM each. The EOMs are interconnected at the fibers between preparation and ultra-stable bench. In this constellation, every single phasemeter will read out one beatnote. Additionally, the GHz clocktone used for one single phasemeter will be imprinted together with ranging and data communication via the EOMs to one of the laser carriers. To assure a non-DC beanote signal between sidebands, slightly offset GHz frequencies will be used. The whole setup can be seen in Fig. 2. In this final version, the three mentioned individual assemblies can be interpreted as LISA satellites. In contrast to other testbeds set up to test primarily the TDI algorithm [17–19], the LIME (LISA metrology experiment) presented here is designed to test and analyze the whole LISA metrology chain from the precise phase readout up to the auxiliary functions in great detail.

5. Conclusions & Outlook

The LISA phasemeter implements phase readout and auxiliary function for the LISA metrology chain. A breadboard model of a LISA phasemeter developed in the scope of an ESA contract was successfully tested for phase readout performance in an electrical null measurement. This technique however cannot reveal non-linearities common in the used channels. For this purpose, an advanced testbed is under development. It utilizes an ultra-stable hexagonal optical bench to implement an optical three signal test. Besides the phase readout, the whole LISA metrology chain including the auxiliary functions can be tested. The well-tested phasemeter itself could then not only be used for LISA, but also in similar applications, like future geodesy missions. In future work, the testbed would serve as a reference and parts can be replaced with corresponding engineering or flight models to be verified.

Acknowledgments

This work was financially supported by Deutsches Zentrum für Luft- und Raumfahrt (reference 50 OQ 1301) with funding from Bundesministerium für Wirtschaft und Technologie.

References

[1] Danzmann K and Rüdiger A 2003 *Class. Quantum Grav.* **20** S1
[2] Danzmann K *et al.* 2015 *Nature Phys.* **11** 613–5
[3] Jennrich O 2009 *Class. Quantum Grav.* **26** 153001
[4] Pierce R, Leitch J, Stephens M, Bender P and Nerem R 2008 *Appl. Opt.* **47** 5007–19
[5] Shaddock D, Ware B, Halverson P G, Spero R E and Klipstein B 2006 *AIP Conf. Proc.* **873** 654–60
[6] Gerberding O, Sheard B, Bykov I, Kullmann J, Delgado J J E, Danzmann K and Heinzel G 2013 *Class. Quantum Grav.* **30** 235029
[7] Hellings R W 2001 *Phys. Rev. D* **64**(2) 022002
[8] Tinto M, Estabrook F B and Armstrong J W 2002 *Phys. Rev. D* **65**(8) 082003
[9] Barke S, Tröbs M, Sheard B, Heinzel G and Danzmann K 2010 *Appl. Phys. B* **98** 33–39
[10] Tinto M and Yu N 2015 *Phys. Rev. D* **92**(4) 042002
[11] Tinto M and Armstrong J W 1999 *Phys. Rev. D* **59**(10) 102003
[12] Otto M, Heinzel G and Danzmann K 2012 *Class. Quantum Grav.* **29** 205003
[13] Sutton A, McKenzie K, Ware B and Shaddock D A 2010 *Opt. Express* **18** 20759–73
[14] Esteban J J, García A F, Barke S, Peinado A M, Cervantes F G, Bykov I, Heinzel G and Danzmann K 2011 *Opt. Express* **19** 15937–46
[15] Gerberding O *et al.* 2012 *ASP Conf. Series* **467** 271–6
[16] Gerberding O *et al.* 2015 *Rev. Sci. Instrum.* **86** 074501
[17] Spero R *et al.* 2011 *Class. Quantum Grav.* **28** 094007
[18] Mitryk S J, Mueller G and Sanjuan J 2012 *Phys. Rev. D* **86**(12) 122006
[19] Gruning P, Halloin H, Brossard J, Prat P, Baron S, Buy C and Jimenez P 2015 *J. Phys.: Conf. Series* **610** 012035

Measuring Earth: Current status of the GRACE Follow-On Laser Ranging Interferometer

Daniel Schütze ***on behalf of the LRI team***

Max Planck Institute for Gravitational Physics (Albert Einstein Institute) and Institute for Gravitational Physics, Leibniz University Hannover, Callinstr. 38, D-30167 Hannover, Germany

Daniel.Schuetze@aei.mpg.de

Abstract. The GRACE mission that was launched in 2002 has impressively proven the feasibility of low-orbit satellite-to-satellite tracking for Earth gravity observations. Especially mass transport related to Earth's hydrological system could be well resolved both spatially and temporally. This allows to study processes such as polar ice sheet decline and ground water depletion in great detail. Owing to GRACE's success, NASA and GFZ will launch the successor mission GRACE Follow-On in 2017. In addition to the microwave ranging system, GRACE Follow-On will be the first mission to use a Laser Ranging Interferometer as technology demonstrator to track intersatellite distance changes with unprecedented precision. This new ranging device inherits some of the technologies which have been developed for the future spaceborne gravitational wave detector LISA. I will present the architecture of the Laser Ranging Interferometer, point out similarities and differences to LISA, and conclude with the current status of the flight hardware production.

Figure 1. Artistic view of the GRACE Follow-On satellites. The microwave beam is shown in blue, the laser beams in red. Watch a brief animation about GRACE Follow-On and the Laser Ranging Interferometer by scanning the QR code or by going to www.gracefo.spacegravity.org/video1. Image credit: Satellites: AEI/Daniel Schütze, Earth: NASA

1. Introduction: Measuring Earth

The advent of modern satellite geodesy made it possible to monitor spatial and temporal variations in Earth's global gravity field [1-3]. Especially the *Gravity Recovery and Climate Experiment* (GRACE, [3]) mission has proven that satellite geodesy is an outstanding tool to investigate processes that are

Journal of Physics: Conference Series **716** (2016) 012005 doi:10.1088/1742-6596/716/1/012005

vital to humankind: By sensing hydrological mass redistribution on a monthly scale, GRACE enables us to investigate climate change induced polar ice mass loss and human impact on ground water depletion [4-5]. A very nice online tool provided by the *Centre National d'Etude Spatiales* (CNES) and the *Groupe des Recherches de Géodésie Spatiale* (GRGS) at www.thegraceplotter.com illustrates the capabilities of GRACE's measurements.

The two identical GRACE satellites were launched in 2002 as a joint mission of the *National Aeronautics and Space Agency* (NASA) and the *Helmholtz Centre Potsdam – German Research Centre for Geosciences* (GFZ). They circle Earth in a common, freely decaying, low polar orbit of currently 380 kilometers altitude with an intersatellite separation of 170–270 kilometers, which is maintained by occasional orbit maneuvers. See the artistic impression in Figure 1.

Spatial variations in Earth's gravity field exert differential accelerations on the two satellites, which leads to length fluctuations in the intersatellite separation. These length fluctuations are tracked with micrometer precision with a microwave ranging device. In conjunction with *Global Positioning System* (GPS) measurements, the underlying gravity field of Earth can be derived [6]. Non-gravitational forces acting on the two satellites, such as atmospheric drag and solar radiation pressure from the sun and Earth's albedo, are removed from the measurements using an accelerometer [7]. After one month, the ground track of the satellites provides sufficient coverage of Earth to yield an update of the gravity field.

GRACE has already greatly exceeded its initially allocated lifetime of 5 years. Battery fatigue on both satellites limits the mission operation time until 2017 to early 2018. Owing to GRACE's great success and the urgent need to continue the observation of Earth's gravity field, NASA and GFZ are currently preparing the GRACE Follow-On satellites, which will be launched in 2017.

While the main science instrument on GRACE Follow-On will still be a microwave ranging device, the two satellites will additionally carry a Laser Ranging Interferometer (LRI, [8]) as a technology demonstrator. This novel instrument will greatly improve the intersatellite range measurements by more than one order of magnitude and additionally deliver highly accurate beam pointing information. Moreover, the LRI will prove the feasibility of interferometric laser ranging technology for next-generation geodesy missions based on high-precision intersatellite ranging.

2. GRACE Follow-On and the Laser Ranging Interferometer

2.1. The LRI project team

The LRI onboard the GRACE Follow-On satellites is a joint project with US and German contributions. On the US side, developments are carried out by NASA and the *Jet Propulsion Laboratory* (JPL), with major components being delivered by *Ball Aerospace* and *Tesat-Spacecom* GmbH & Co. KG. The US contributions include the phasemeter and payload processing electronics, the laser frequency stabilization [9], and the laser source.

The German side of the project team consists of GFZ, the *Max Planck Institute for Gravitational Physics* (*Albert Einstein Institute*, AEI), the *Deutsches Zentrum für Luft- und Raumfahrt* (DLR), *SpaceTech* GmbH, and *Airbus Defence & Space* GmbH, SAS, SA & Ltd. The German responsibilities are the optics, the Triple Mirror Assembly, the beam steering method including electronics, and the quadrant photo receivers with electronics. According to the cooperative project plan, the AEI provides the German LRI instrument manager and STI the German LRI implementation manager.

Additional support for the LRI project was provided by the *Australian National University* (ANU), *Electro Optic Systems* Pty. Ltd., and the *Commonwealth Scientific and Industrial Research Organisation* (CSIRO), where the first Triple Mirror Assembly prototypes were built and tested [10, 11].

2.2. Comparison between LISA and GRACE Follow-On laser ranging

Advanced interferometric intersatellite ranging technologies have been developed for the *Laser Interferometer Space Antenna* (LISA, [12]), which is going to observe low-frequency gravitational waves in the millihertz regime. Since GRACE Follow-On relies on high-precision intersatellite ranging as well, it is worthwhile to consider relevant similarities and differences between the laser ranging technologies applied in the respective mission scenarios, which are listed in Table 1.

The baseline ranging technology for both LISA and the GRACE Follow-On LRI is heterodyne laser interferometry with frequency offset phase-locked lasers. In both cases, relative intersatellite line-of-sight velocities are several meters per second, so that an offset frequency of several megahertz has to be applied to keep the ranging signal unambiguous under changing Doppler shifts.

Although the intersatellite separation is quite different for LISA and GRACE Follow-On, being 1 Mio. kilometers and 200 kilometers, respectively, the received light power levels of 100 picowatts are quite similar. This is due to the fact that GRACE Follow-On uses much lower laser powers to start with and employs no telescope for weight and space constraint reasons.

A major difference between LISA and GRACE Follow-On can be found in the orbit environmental conditions. While LISA will be launched into a heliocentric orbit with very stable environmental conditions, GRACE Follow-On is operating in a low-Earth orbit with considerable atmospheric drag and changing solar radiation. This requires an active beam steering scheme to counteract spacecraft attitude jitter.

Table 1. Similarities and differences between LISA and GRACE Follow-On laser ranging technologies.

	LISA	**GRACE Follow-On**
Intersatellite distance measurement	Heterodyne laser interferometry (receiver-transponder) with offset phase-locked lasers and phasemeter readout	
Intersatellite distance	1.000.000 km	• 200 km
Orbit	Heliocentric (1 au)	Polar low-Earth orbit (490 km initial altitude)
Orbit environment	No atmospheric drag, thermally stable	Atmospheric drag, large thermal disturbances
Attitude & orbit control system	Drag-free with μN-thrusters	Attitude control with magneto-torquers & cold gas thrusters
Measurement band	• mHz…Hz	
Single-link noise goal	12 pm/$\sqrt{\text{Hz}}$ (×frequency dependence)	80 nm/$\sqrt{\text{Hz}}$ (×frequency dependence)
Telescope aperture diameter	38 cm	8 mm (no telescope, just clipping by aperture stop)
Transmit beam waist diameter	34 cm	5 mm
Transmit power	1 W	30 mW
Effective received power at photo receiver	• 100 pW	
Maximum relative line-of-sight velocity	±15 m/s	±5 m/s

However, the ranging requirements are much relaxed for GRACE Follow-On compared to LISA, aiming at 80 nm/$\sqrt{\text{Hz}}$, while LISA requires 12 pm/$\sqrt{\text{Hz}}$ for gravitational wave observation. Due to GRACE's orbit period of about 90 minutes, the observation frequency band falls into the millihertz regime as for LISA.

In summary, the LRI for GRACE Follow-On is an adaptation of LISA technology to quite a different mission scenario in terms of environmental noise sources and required ranging precision. This relaxes much of LISA's requirements, however, new challenges arise from the much rougher orbit environment.

2.3. Architecture of the GRACE Follow-On Laser Ranging Interferometer

A schematic of the LRI [8] on one of the two GRACE Follow-On spacecraft is shown in Figure 2. The LRI is identical on both spacecraft, so we can focus our explanations on one spacecraft. For the constellation of the two satellites, keep in mind the artistic view shown in Figure 1. Additionally, this short video clip gives you a better idea of the geometry inside one of the GRACE Follow-On spacecraft: www.gracefo.spacegravity.org/video2.

The LRI uses an active transponder principle, which means that the weak incoming received (RX) beam is replaced by a strong local oscillator (LO) beam. This beam is then retroreflected by the Triple Mirror Assembly (TMA) and sent to the distant spacecraft.

The TMA is a passive retroreflector which consists of three perpendicular mirror sections – for a better visualization, watch the animation at www.gracefo.spacegravity.org/video3. Originally, the TMA was introduced to solve space constraint challenges and to route the laser beam around the cold gas tank and the microwave ranging system, which are blocking the direct axis between the two spacecraft's center-of-masses (CoM). However, it turned out that the LRI architecture using the TMA allows for a simple implementation of closed-loop beam steering, while the TMA's geometrical properties suppress coupling of spacecraft attitude jitter into the pathlength measurement [13].

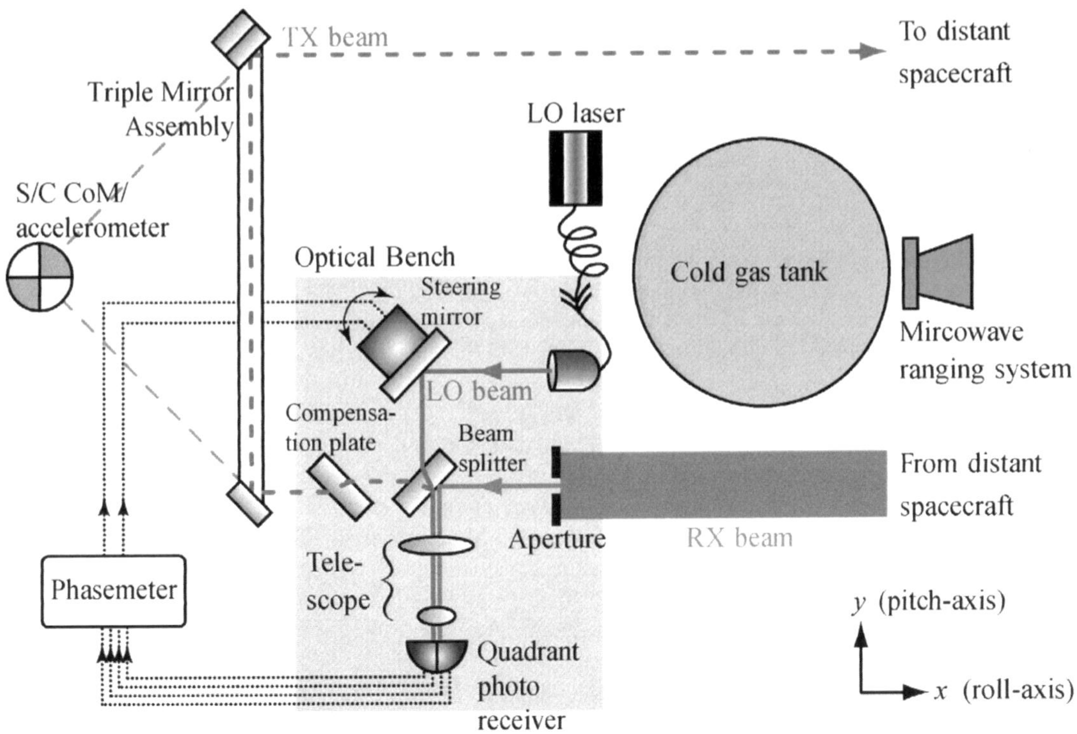

Figure 2. Schematic of the LRI on one of the two GRACE Follow-On satellites. The LRI is identical on both satellites.

The light received from the distant spacecraft (RX beam) is clipped at an 8 millimeter diameter aperture on the Optical Bench. This clipped beam is superimposed with a strong local oscillator (LO) beam and imaged with a telescope on the quadrant photo receiver (QPR).

The four phases of the QPR channels are recorded with a phasemeter. The average phase serves as error signal to drive the offset phase-lock of the LO laser on the slave spacecraft. On the master spacecraft, the average phase contains the intersatellite ranging information. By adding and subtracting the four QPR channels appropriately, signals sensitive to 2-dimensional relative wavefront tilt between LO wavefronts and RX wavefronts can be generated, which is known as Differential Wavefront Sensing (DWS). A control loop is implemented in the phasemeter, which feeds back the DWS signals to a 2-axes steering mirror. By this, local spacecraft attitude jitter, which leads to a tilt of the RX wavefronts with respect to the LO wavefronts, is actively counteracted [14].

3. In three years from prototypes to flight hardware

The LRI is quite a unique space project. As a technology demonstrator, some of the test and redundancy requirements, which are usually applied to space missions, are relaxed. This made it possible for this innovative and ambitious project to evolve from sheer design studies in 2010, to first prototypes and breadboard level verifications in 2012, to construction and test of engineering models thereafter, and ultimately to the construction and test of flight units in 2014/2015. Currently, the last flight subsystems are being assembled and tested. In a next step, the LRI flight units will be integrated into the GRACE Follow-On satellites.

4. Conclusion

The GRACE Follow-On satellites, which will be launched in 2017, will continue GRACE's quest of monitoring spatial and temporal variations in Earth's gravity field. For the first time, the intersatellite range displacements, from which the gravity potential is deduced, will be measured with a Laser Ranging Interferometer. This novel device promises an improved ranging precision compared to the conventional microwave ranging system by more than one order of magnitude to achieve 80 $\mathrm{nm}/\sqrt{\mathrm{Hz}}$ in the measurement band of 0.002…0.1 Hz.

Flight hardware production for the Laser Ranging Interferometer is currently being finished. Most flight hardware has already successfully passed the test procedures. In the next step, the Laser Ranging Interferometer will be integrated into the two GRACE Follow-On spacecraft to be ready for launch in 2017.

Acknowledgments

This work was partly funded by the "Bundesministerium für Bildung und Forschung" (BMBF, project number: 03F0654B) and by the "Deutsche Forschungsgemeinschaft" (DFG) within the Cluster of Excellence QUEST (Centre for Quantum Engineering and Space-Time Research).

References

[1] C. Reigber, H. Luhr, and P. Schwintzer, "CHAMP mission status," Adv. Space Res. **30,** 129–134 (2002).

[2] R. Pail, S. Bruinsma, F. Migliaccio, C. Forste, H. Goiginger, W. D. Schuh, E. Hock, M. Reguzzoni, J. M. Brockmann, O. Abrikosov, M. Veicherts, T. Fecher, R. Mayrhofer, I. Krasbutter, F. Sanso, and C. C. Tscherning, "First GOCE gravity field models derived by three different approaches," J. Geodesy **85,** 819–843 (2011).

[3] B. D. Tapley, S. Bettadpur, M. Watkins, and C. Reigber, "The gravity recovery and climate experiment: mission overview and early results," Geophys. Res. Lett. **31,** L09607 (2004).

[4] I. Velicogna and J. Wahr, "Time-variable gravity observations of ice sheet mass balance: precision and limitations of the GRACE satellite data," Geophys. Res. Lett. **40,** 3055–3063 (2013).

[5] V. M. Tiwari, J. Wahr, and S. Swenson, "Dwindling groundwater resources in northern India, from satellite gravity observations," Geophys. Res. Lett. **36,** L18401 (2009).

[6] C. Dunn, W. Bertiger, Y. Bar-Sever, S. Desai, B. Haines, D. Kuang, G. Franklin, I. Harris, G. Kruizinga, T. Meehan, S. Nandi, D. Nguyen, T. Rogstad, J. B. Thomas, J. Tien, L. Romans, M. Watkins, S. C. Wu, S. Bettadpur, and J. Kim, "Instrument of GRACE: GPS augments gravity measurements," GPS World **14,** 16–28 (2003).

[7] P. Touboul, E. Willemenot, B. Foulon, and V. Josselin, "Accelerometers for CHAMP, GRACE and GOCE space missions: synergy and evolution," B. Geofis. Teor. Appl. **40,** 321–327 (1999).

[8] B. S. Sheard, G. Heinzel, K. Danzmann, D. A. Shaddock, W. M. Klipstein, and W. M. Folkner, "Intersatellite laser ranging instrument for the GRACE follow-on mission," J. Geodesy **86,** 1083–1095 (2012).

[9] W. M. Folkner, G. deVine, W. M. Klipstein, K. McKenzie, D. Shaddock, R. Spero, R. Thompson, D. Wuchenich, N. Yu, M. Stephens, J. Leitch, M. Davis, J. deCino, C. Pace, and R. Pierce, "Laser frequency stabilization for GRACE-II," in *Proceedings of the 2011 Earth Science Technology Forum* (2011).

[10] R. L. Ward, R. Fleddermann, S. Francis, C. Mow-Lowry, D. Wuchenich, M. Elliot, F. Gilles, M. Herding, K. Nicklaus, J. Brown, J. Burke, S. Dligatch, D. Farrant, K. Green, J. Seckold, M. Blundell, R. Brister, C. Smith, K. Danzmann, G. Heinzel, D. Schütze, B. S. Sheard, W. Klipstein, D. E. McClelland, and D. A. Shaddock, "The design and construction of a prototype lateral-transfer retro-reflector for inter-satellite laser ranging," Class. Quantum Grav. **31,** 095015 (2014).

[11] D. Schütze, D. Farrant, D. A. Shaddock, B. S. Sheard, G. Heinzel, and K. Danzmann, "Measuring coalignment of retroreflectors with large lateral incoming-outgoing beam offset," Rev. Sci. Instrum. **85,** 035103 (2014).

[12] K. Danzmann *et al.*, "The gravitational universe," European Space Agency (ESA) L2/L3 Science Theme Selection, arXiv:1305.5720 (2013).

[13] D. Schütze, V. Müller, G. Stede, B. S. Sheard, G. Heinzel, K. Danzmann, A. J. Sutton, and D. A. Shaddock, "Retroreflector for GRACE Follow-On: Vertex vs. point of minimal coupling," Opt. Express **22,** 9324–9333 (2014).

[14] D. Schütze, G. Stede, V. Müller, O. Gerberding, T. Bandikova, B. S. Sheard, G. Heinzel, and K. Danzmann, "Laser beam steering for GRACE Follow-On intersatellite interferometry," Opt. Express **22,** 24117–24132 (2014).

How beaming of gravitational waves compares to the beaming of electromagnetic waves: impacts to gravitational wave detection

Andrew L. Miller[1] **and Thulsi Wickramasinghe**[1]

[1] Department of Physics, The College of New Jersey, 2000 Pennington Road, Ewing, NJ 08628, USA

E-mail: `millea12@tcnj.edu`

Abstract. We focus on understanding the beaming of gravitational radiation from gamma ray bursts (GRBs) by approximating GRBs as linearly accelerated point masses. For accelerated point masses, it is known that gravitational radiation is beamed isotropicly at high speeds, and beamed along the polar axis at low speeds. Aside from this knowledge, there has been very little work done on beaming of gravitational radiation from GRBs, and the impact beaming could have on gravitational wave (GW) detection. We determine the following: (1) the observation angle at which the most power is emitted as a function of speed, (2) the maximum ratio of power radiated away as a function of speed, and (3) the angular distribution of power ratios at relativistic and non-relativistic speeds. Additionally the dependence of the beaming of GW radiation on speed is essentially the opposite of the beaming of electromagnetic (EM) radiation from GRBs. So we investigate why this is the case by calculating the angular EM radiation distribution from a linear electric quadrupole, and compare this distribution to the angular gravitational radiation distribution from a GRB.

1. Introduction

Binary neutron stars, binary black holes and neutron star-black hole pairs are the most likely source of gravitational waves. Emissions from these compact binaries are expected to be short gamma ray bursts (GRBs), which are very short bursts of large amounts of electromagnetic energy. GRBs are divided into two types based on the signal's duration and spectral hardness: "short" and "long" GRBs[1]. The origin of "short" GRBs is thought to be from compact binaries coalescence (CBC). If gravitational waves are detected by advanced LIGO from CBCs, we can further distinguish between the signals produced by different binary systems. "Long" GRBs mostly originate from stellar collapses and are not that relevant here. An overly simplistic equation used to compute the so-called "isotropic" energy due to a GRB is given as[2]:

$$E_{iso} = \frac{4\pi(BC){D_L}^2 f}{1+z} \tag{1}$$

where E_{iso} refers to the energy due to the GRB that is emitted in all directions equally, z is the redshift, D_L is the luminosity distance to the source, f is the fluence measured in the High Energy Transient Explorer (HETE-2) waveband, and BC is the bolometric correction for this waveband for GRBs. E_{iso} is taken to be $10^{-2}M_{sun}c^2$. However it is not certain that the

CBC sources in LIGO's sensitivity range will emit GRBs isotropically. In fact if the bursts are at lower speeds, they will be beamed along the polar axis, which might present a better chance of detection. However it has been determined that if the GWs are beamed along the polar axis, the power due to these signals would have been underestimated by the factor u:

$$u = 1 + \frac{11x}{16} + \frac{11x^2}{16} + \frac{x^3}{16} + \frac{x^4}{16} \tag{2}$$

where $x = \cos\psi$ and ψ is the opening angle around the normal into which the bursts are beamed[3].

And if GRBs are emitted isotropicly, they might not be strong enough to be detected, since E_{iso} varies among order of magnitudes for different known GRBs[4]. So we need to determine what proportion of this variable amount of power is actually detectable.

Additionally, we know that gravitational waves are the result of a time-varying quadrupole moment of a source, but there exists a fundamental difference in the way that electromagnetic radiation from an accelerated point charge behaves at relativistic speeds compared to gravitational radiation from an accelerated point mass at relativistic speeds:

$$\frac{dE}{d\Omega d\omega} = \frac{e^2}{4\pi^2 c}\beta^2\gamma^2\frac{\sin^2\theta}{(1-\beta\cos\theta)^2} \tag{3}$$

$$\frac{dE}{d\Omega d\omega} = \frac{Gm^2}{4\pi c^2}\beta^4\gamma^2\frac{\sin^4\theta}{(1-\beta\cos\theta)^2} \tag{4}$$

Here $\frac{dE}{d\Omega d\omega}$ is the energy emitted per solid angle Ω per angular frequency ω, θ is the observation angle, and the rest of the variables carry their usual meanings. At relativistic speeds, equations 3 and 4 indicate that EM radiation is beamed and that GW radiation is not (it is in fact isotropic). We investigate why this fundamental difference occurs by accelerating a linear electric quadrupole (two dipoles back-to-back) and calculating its power distribution. There has been work done on calculating the electric and magnetic fields of a dipole in motion, and the corresponding power distributions as a function of θ, the observation angle[5]. In this case, the power distribution of a linearly accelerating dipole is even more beamed than for an accelerated point charge. However, no such calculation has been done for a linear electric quadrupole in motion, nor has an angular power distribution ever been computed.

2. Methods

Even though GRBs come from in-spiraling binaries, we approximate GRBs as accelerated point masses, whose gravitational power distribution is given by equation 4. We ask the following questions: (1) at what observation angle do we receive the greatest ratio of power, (2) how is this ratio of power affected by speed of the source, and (3) given a relativistic speed, what will be the source's angular distribution of power?

In order to answer question (1), we differentiate equation 4 with respect to θ, and determine the angle that corresponds to a maximum in emitted power. We vary the speed β to see how the maximum angle changes as the source moves more and more relativistically. For question (2), we integrate equation 4 with respect to θ to determine the total power emitted over all solid angles. After, we compute the ratio of maximum power to total power for a given source and vary the speed to determine the impact of changing speeds on the ratio of power emitted. For question (3), we plot the ratio of power at a given speed to the total power (integrated over all angles) as a function of the angle θ.

To address the fundamental difference between gravitational and electromagnetic power angular distributions, we find the angular power distribution due to a accelerated electric quadrupole, and see if that is similar to the angular power distribution given in equation 4. For simplicity, we accelerate a linear electric quadrupole in one direction, as shown in figure 1.

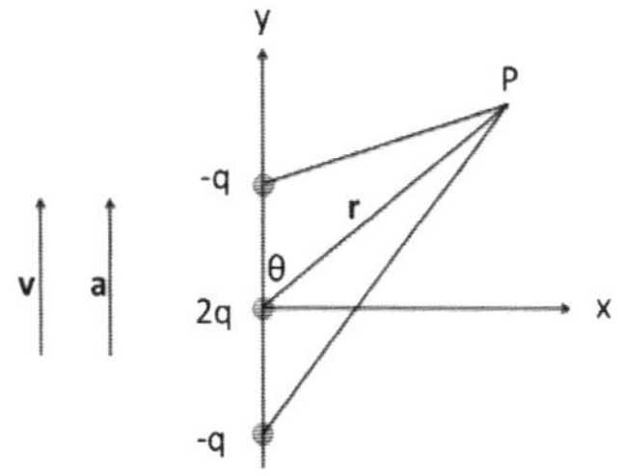

Figure 1. Model of accelerating linear electric quadrupole.

We compute the angular power distribution at point P at a distance r from the center of the linear electric quadrupole, by (1) calculating the retarded scalar and vector potentials V and **A** from the Lienard-Wiechert potentials, (2) finding **E** and **B** (keep only 1/r terms), (3) computing the Poynting vector **S**, and (4) finding the angular power distribution from **S**, with $\frac{dP}{d\Omega} = |\mathbf{S}|r^2$.

For simplicity, we assume that the distances from all the charges to the point P are equal to r, and that the angles that the negative charges make with the normal are also θ. In doing this, if we see similar angular power distributions between an accelerated linear electric quadrupole and a linearly accelerated point mass, then we can conclude that the "changing quadrupole" nature of the GW may be what causes this angular distribution of power.

3. Results

We determine the observation angle per unit solid angle that corresponds to the greatest ratio of power emitted as a function of speed for an accelerated point mass, which is shown in figure 2a. Figure 2b shows how the ratio of power corresponding to the maximum observation angle changes as a function of speed. We also plot the angular distribution of power for a given speed, to show where we need to look to see the most power. We see symmetric maxima at $\theta \approx 48°$ and a local maximum at $\theta = \pi/2$. See figure 3.

We calculated the leading term of the angular power distribution for an accelerated linear electric quadrupole to be:

$$\frac{dP}{d\Omega} \propto a^2 q^4 \frac{\beta^2 \cos^2\theta}{(1-\beta^2\cos^2\theta)^6} \tag{5}$$

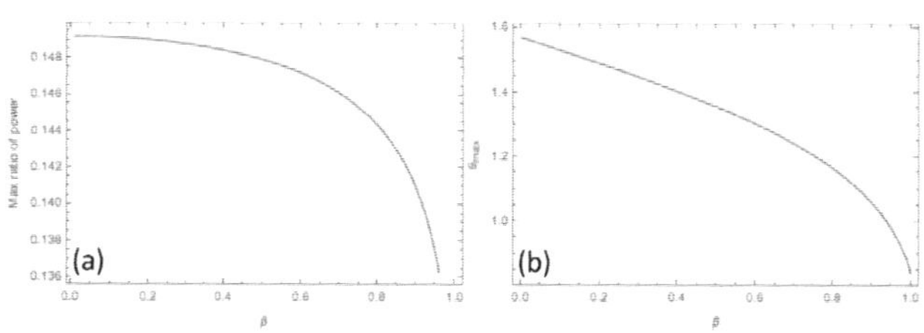

Figure 2. (a) Ratio of power emitted at max angle vs. speed, for linearly accelerated point mass, (b) Angle at which max power is emitted vs. speed.

There are other terms in equation 5 with higher-orders of β and $\cos\theta$. However, when neglected, they do not change the angular behavior at high speeds. See figure 4. At lower speeds, the distribution is relatively isotropic, as shown in figure 5, and the lower-order terms cause a downward "bending" of the isotropic distribution.

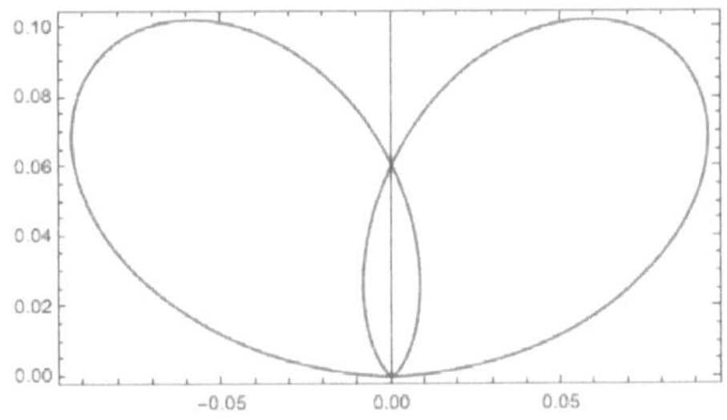

Figure 3. Angular power distribution for $\beta = 0.999$ for linearly accelerated point mass.

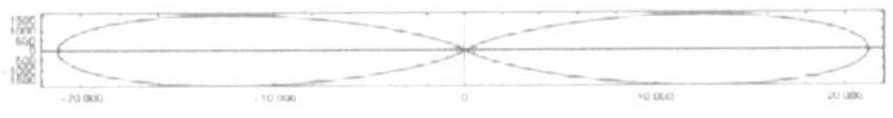

Figure 4. Accelerated electric quadrupole power distribution, leading order term, $\beta = 0.9$.

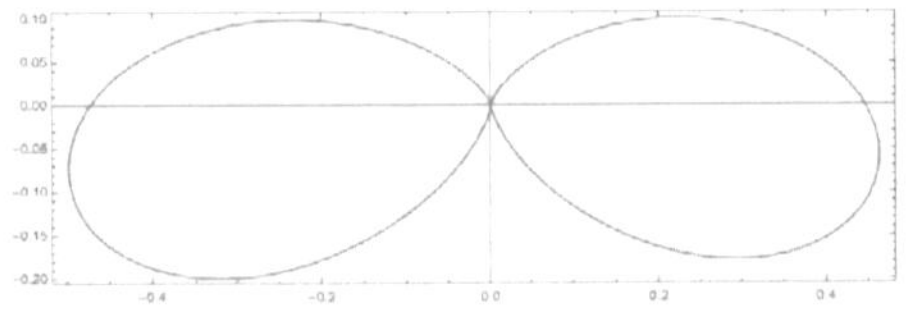

Figure 5. Accelerated electric quadrupole power distribution, $\beta = 0.1$.

4. Discussion

Based on figure 2, only a small fraction (12%-15%) of power from a GRB per unit solid angle is received by us even if we are observing it from the optimal angle. Additionally a GRB emitted directly at us at $\beta = 0.999$ will not yield the greatest ratio of power ($\theta_{max} \approx 48°$). We have also seen that the angle at which we will see the most power decreases at higher speeds. Gravitational wave detection is hindered by the angular distribution of accelerated point masses: we therefore believe that radiation from bursts will be harder to see in most scenarios. Even in the best case scenario (at the optimal viewing angle), only around 15% of the total power per unit solid angle will reach us. The event will need to be very strong and emitted at the optimal angle to be seen.

Figures 4 and 5 show that the quadrupole moment is probably not responsible for the differing EM and GW angular power distributions. This is because the EM radiation due to an accelerating electric quadrupole at relativistic speeds is beamed along the observation axis, and the radiation is isotropic at lower speeds. Gravitational radiation from an accelerated point mass behaves exactly opposite to this.

We recognize that a better quadrupole to use would have been a square of charge in two dimensions centered at the origin, since it does not have the higher-order multipoles that a linear electric quadrupole does. We think this calculation would yield more insight into the reasons why changing quadrupole moments drive gravitational wave emission.

5. References

[1] Nakar E 2007 *Phys.Rept.* **442** 166-236
[2] Abbott B et al. 2005 *Phys. Rev. D* **72** 042002
[3] Kochanek CS and Piran T 1993 *Astrophys.J.* **417** L17-L20
[4] Frail DA et al. 2001 *ApJ* **562** L55
[5] Mott H, Hoadley GB, and Davis WR 1967 *Journal of the Franklin Institute*, **283(2)** 91-101.

Tackling excess noise from bilinear and nonlinear couplings in gravitational-wave interferometers

Sukanta Bose[1,2], Bernard Hall[1], Nairwita Mazumder[1], Sanjeev Dhurandhar[2], Anuradha Gupta[2], Andrew Lundgren[3]

[1] Department of Physics & Astronomy, Washington State University, 1245 Webster, Pullman, WA 99164-2814, U.S.A

[2] Inter-University Centre for Astronomy and Astrophysics, Post Bag 4, Ganeshkhind, Pune 411 007, India

[3] Max-Planck-Institut für Gravitationsphysik (Albert-Einstein-Institut), Callinstrasse 38, 30167 Hannover, Germany

E-mail: sukanta@wsu.edu

Abstract.
We describe a tool we improved to detect excess noise in the gravitational wave (GW) channel arising from its bilinear or nonlinear coupling with fluctuations of various components of a GW interferometer and its environment. We also describe a higher-order statistics tool we developed to characterize these couplings, e.g., by unraveling the frequencies of the fluctuations contributing to such noise, and demonstrate its utility by applying it to understand nonlinear couplings in Advanced LIGO engineering data. Once such noise is detected, it is highly desirable to remove it or correct for it. Such action in the past has been shown to improve the sensitivity of the instrument in searches of astrophysical signals. If this is not possible, then steps must be taken to mitigate its influence, e.g., by characterizing its effect on astrophysical searches. We illustrate this through a study of the effect of transient sine-Gaussian noise artifacts on a compact binary coalescence template bank.

1. Introduction

The advanced detector era was ushered in by the two Advanced LIGO (aLIGO) detectors [1] in September 2016. With plans afoot for the Advanced Virgo (AdV) detector [2] to join it this decade, we expect the era of gravitational wave (GW) astronomy to begin in earnest [3]. The sensitivity of an interferometer in the detection band (which in aLIGO is currently between a few tens of Hertz to a few kiloHertz) [1], however, can be affected adversely due to unwanted noise arising from various sources. The effect of noise transients (see, e.g., Ref. [4] and references therein) on astrophysical searches was studied in Ref. [5] for binary neutron signals, in Refs. [6, 7] for intermediate mass black hole signals, in Ref. [8] for binary black hole signals, and in Ref. [9] for unmodeled transient signals. Detector characterization efforts that have been invested toward improving the quality of GW data in the recent past are described in Refs. [10, 11, 12]. In this work we improve upon a subset of those efforts, specifically, in regards to identifying the presence of certain types of noise that appear in the detection band owing to bilinear or nonlinear coupling of fluctuations at lower frequencies. We also describe a higher-order statistics tool we developed to characterize these couplings by finding the frequencies of the fluctuations that contribute to such noise. We demonstrate its utility by applying it to aLIGO engineering data.

One of the simplest methods used for identifying the source of excess noise in the GW strain data of a detector is to measure its coherence with data recorded by, say, a sensor monitoring an environmental disturbance or a fluctuation in an interferometer component. Sensors record such data as time-series in multiple channels. For instance, time variations in the pitch, yaw, and roll (angular) degrees of freedom of each test mass can be extracted from the time-series data acquired by the Optical Sensors and Electromagnetic Actuators (OSEMs) attached to the multiple stages and levels of their suspension and recorded in various channels. The power-line harmonics can be extracted from the data recorded by the magnetometers. On the other hand, an example of an environmental disturbance is ground motion, which is measured by seismometers and accelerometers.

The coherence statistic is useful in identifying noise sources, be they internal to the interferometer or environmental, that couple linearly with the GW channel. There are various tools available to find them effectively and, therefore, will not be studied here (see, e.g., Ref. [14] and references therein). Unwanted noise can also arise in the GW channel owing to higher order couplings among multiple channels. For instance, when a servo-control system drifts near the edge of its actuation range, a further small fluctuation can drive the actuator into a nonlinear regime, causing elevated rate of noise transients in the GW channel. This can involve multiple channels in a nonlinear fashion, such as when a large alignment drift reduces the actuation range for additional length fluctuations. Another example would be a bilinear combination of a slow angular drift in alignment that causes a measured light beam to drift off its detector, thus amplifying the effect of any further higher-frequency jitter of the light-beam alignment. This type of behavior increases the noise background of searches for short-duration astrophysical signals, e.g., from binary black hole coalescences, hence, reducing the detector's sensitivity to them [10].

The Bilinear Coupling Veto (BCV) tool [16] has been shown to be useful in identifying the presence of non-Gaussian, bilinear component of the noise to a certain degree. While there are many tools that can identify the presence of component frequencies in a given time series, those typically employed are not able to provide any definitive evidence of nonlinear couplings that may be present (see, however, Ref. [17]). For example, the noise power spectral density (PSD) of the GW channel can show frequencies at which excess noise due to such couplings is present (see, e.g., the side-bands of the violin mode of a LIGO test mass in Fig. 1). However, there is no phase information present here, and one needs to employ higher-order statistics, in general, for diagnosing the presence of non-Gaussianity and nonlinear couplings [18]. BCV addresses this problem, partially, by finding strong coherence between the GW time-series, on the one hand, and the product of the time-series of two interferometer channels, on the other hand. Reference [16] termed this product as a *pseudo* channel. (Note that the pseudo channel can be constructed from environmental channels as well.) Typically, one of the time-series in that product is taken from a channel that records slow variation, e.g., the changing pitch-angle of an end-test mass, and the other time-series is taken from one that records fast variation, e.g., the signal controlling the length of the power-recycling cavity. We call this coherence the BCV statistic. In this work, we use the higher-order statistics tool to unearth some characteristics of the noise couplings that BCV finds.

2. Detecting bilinear and nonlinear noise couplings

The last published work on BCV [16] presented results obtained from its application to LIGO's sixth Science Run "S6" data, when the wavelet-based Kleine Welle [19] excess-power detection tool was used to identify noise transients. In preparation for aLIGO observations, we have upgraded BCV to work (a) additionally with transients found by "Omicron", which uses the Q-transform on a sine-Gaussian basis [20, 19], and (b) with aLIGO channels, which use a different nomenclature and are more numerous than LIGO channels.

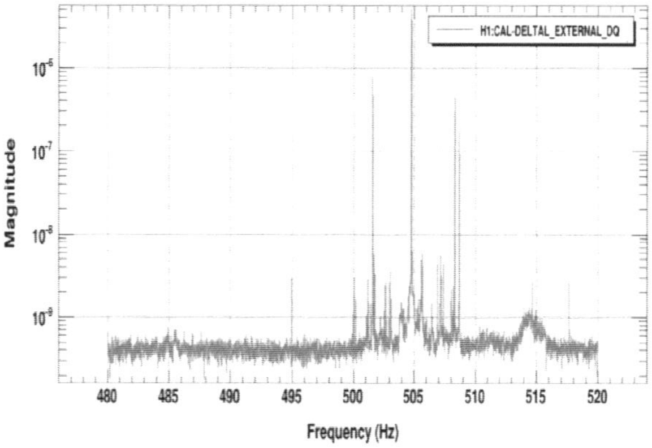

Figure 1. Excess noise in the form of side-bands of a violin mode, at approximately 505 Hz, suggests the possible existence of nonlinear coupling between the strain data in the GW channel (which is represented here by "H1:CAL-DELTAL_EXTERNAL_DQ") and noise in a host of other channels. (The discussion around Eq. (3) explains the origin of such side-bands.)

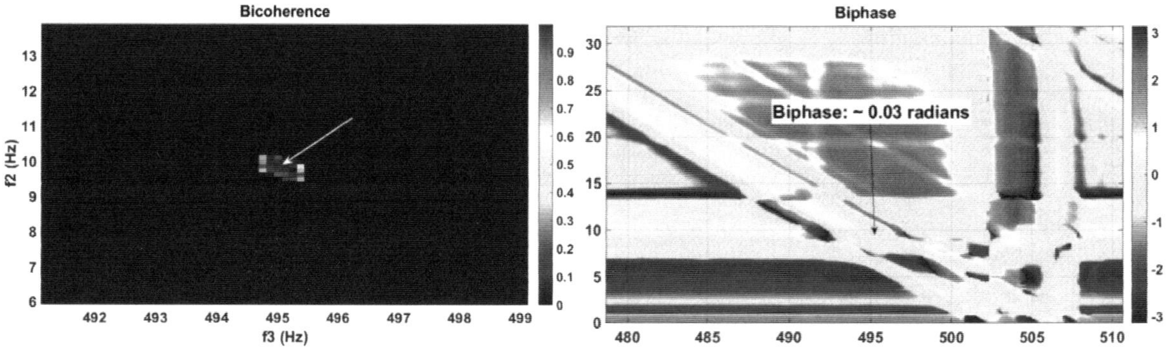

Figure 2. The bounce mode (f_2 ≈10 Hz) of one of the test masses couples with its violin mode (f_1 ≈505 Hz) to create a side-band at (f_3 ≈495 Hz). The color-bar shows the bicoherence and the biphase (in radians) in the left and right figures, respectively, computed on the GW strain data.

We use BCV to identify episodes of bilinear coupling as follows. First, the occurrence times and other characteristics of excess noise transients or "glitchiness" are identified by Omicron or Kleine Welle in the GW channel and multiple instrumental and environmental channels that record fast variations. Second, time-coincidences of transients in that latter set of channels and the GW channel are identified for possible correlations between them. Third, during times of coincidences the BCV statistic is computed on the GW time-series and a pseudo channel, which is a product of the time-series with fast variations and a time-series picked from channels recording slow variations. For every coincidence, the last step is repeated over multiple channels that record slow variations. A strong value of the BCV statistic indicates the presence of bilinear coupling. To assess how large the value of this statistic can be in the absence of a coupling, the BCV statistic is computed also for acausally large relative time-shifts between the GW channel and the pseudo channel of interest. (Reference [16] used the bilinear coherence statistic for identifying bilinear coupling, but it is possible to replace it with the *bicoherence* statistic defined below for diagnosing the presence of nonlinear couplings.) When the BCV statistic is found

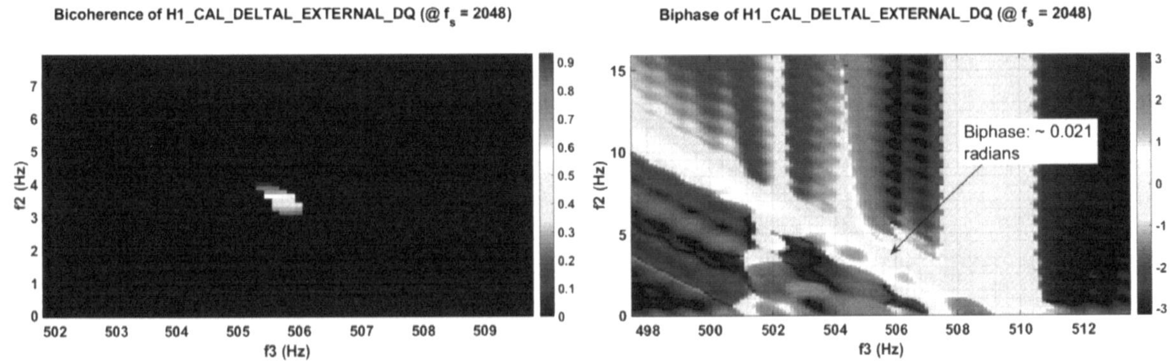

Figure 3. A mechanical resonance at ($f_2 \approx$3.8 Hz) of the suspension of one of the test masses couples with its violin mode ($f_1 \approx$509 Hz) to create a side-band at ($f_3 \approx$505.2 Hz). The color-bar shows the bicoherence and the biphase (in radians) in the left and right figures, respectively, computed on the GW strain data.

to be significantly larger than the background, the GW data is flagged for further data quality studies and for setting vetoes for astrophysical searches. However, to search for the sources of these transients, one must unearth more information about their characteristics than just what channels are involved in producing such bilinear / multi-linear or nonlinear couplings. For example, the frequencies of the disturbances that couple provide valuable information about the source and pathways of such couplings. Below we describe a few of those statistics that we use to find such information.

3. Characterizing noise couplings by employing higher order statistics

One possible way of probing the nature of beyond-linear-order couplings is to compute a higher order statistic, e.g., the bicoherence, of the disturbances that we suspect can couple with one another. Strong bicoherence between channels can suggest couplings of their respective subsystems. For instance, mechanical resonances of structures (e.g., periscopes) show up at higher frequencies in the GW and auxiliary channels as well as some sensor channels.

Mathematically, a set of N disturbances or "signals" x_k is said to combine linearly when they superimpose to form a new signal, $z_{\rm lin}$, as follows:

$$z_{\rm lin}(t) = \sum_{k=1}^{N} x_k(t)\,, \tag{1}$$

In the special case where the x_k are sinusoidal, one gets

$$z_{\rm lin}(t) = \sum_{k=1}^{N} a_k \cos(2\pi f_k t + \phi_k)\,, \tag{2}$$

where a_k and f_k are the amplitudes and frequencies of the linearly combining signals, and ϕ_k are their initial phases, respectively. When $N = 2$, we will identify $x_1(t)$ as $x(t)$ and $x_2(t)$ as $y(t)$, and write $z_{\rm lin}(t) = x(t) + y(t)$.

Though there are different means by which signals may combine nonlinearly, we will consider one of the simplest yet ubiquitous coupling, namely, a type of coupling known as the quadratic phase coupling (QPC) [21]. The reason for this choice is that it is the most significant term in many nonlinear couplings and is also straightforward to model. In this case, two (or more) signals combine to form a new signal, $z(t)$. For a pair of coupled signals one gets,

$$\begin{aligned} z(t) &= p[x(t)+y(t)]^2 + q[x(t)+y(t)] \\ &= p\left[a_1\cos(2\pi f_1 t+\phi_1)+a_2\cos(2\pi f_2 t+\phi_2)\right]^2 \\ &\quad +q\left[a_1\cos(2\pi f_1 t+\phi_1)+a_2\cos(2\pi f_2 t+\phi_2)\right] , \end{aligned} \quad (3)$$

where p and q are real numbers, and the pair of frequencies, (f_1, f_2), is known as a bifrequency. Heretofore, t will denote a time index. Since the right-hand side has terms that denote oscillations (and power) at frequencies $(f_1 \pm f_2)$, $2f_1$, and $2f_2$, in order to detect QPC it is necessary to use a statistic that can aid the detection of these frequency components. In some of the examples studied in this paper, f_2 is much smaller than f_1. In such an event, excess noise will appear in the sidebands $(f_1 \pm f_2)$ of f_1.

To obtain one such statistic, we first divide the time-series, $x(t)$, $y(t)$, and $z(t)$ into M segments $x_\alpha(t)$, $y_\alpha(t)$, and $z_\alpha(t)$, respectively, where α is the segment index. Next the Fourier transforms, $X_\alpha(f)$, $Y_\alpha(f)$, $Z_\alpha(f)$, of each of these segments is taken, respectively, followed by the computation of the average of the quantity $X_\alpha(f_1)Y_\alpha(f_2)Z^*_\alpha(f_1+f_2)$ over those segments:

$$B(f_1, f_2) = \frac{1}{M}\sum_{\alpha=1}^{M} X_\alpha(f_1)Y_\alpha(f_2)Z^*_\alpha(f_1+f_2) . \quad (4)$$

A normalized and real function derived from $B(f_1, f_2)$ is defined as

$$b(f_1, f_2) = |B(f_1, f_2)| \left[\frac{1}{M}\left|\sum_{\alpha=1}^{M} X_\alpha(f_1)Y_\alpha(f_2)\right| \frac{1}{M}\left|\sum_{\alpha=1}^{M} Z^*_\alpha(f_1+f_2)\right|\right]^{-1} , \quad (5)$$

which takes values in the range $[0, 1]$. It attains large values when p is large, i.e., when a strong QPC is present. Another related statistic, defined for individual segments, is

$$\Delta\phi_\alpha(f_1, f_2) = \arg\left[X_\alpha(f_1)Y_\alpha(f_2)Z^*_\alpha(f_1+f_2)\right] . \quad (6)$$

Its average, $\Delta\phi$, for a large enough M, should be at the level of the noise when a QPC is present. Otherwise it can be larger, depending on the type of oscillations present. The statistic $B(f_1, f_2)$ is related to the bispectrum, $B_Z(f_1, f_2)$, which is obtained by replacing $X_\alpha(f_1)$ and $Y_\alpha(f_2)$ in $B(f_1, f_2)$ by $Z_\alpha(f_1)$ and $Z_\alpha(f_2)$, respectively. The same replacements in $b(f_1, f_2)$ and $\Delta\phi$ yield the bicoherence, $b_Z(f_1, f_2)$ and biphase $\Delta\phi_Z$. Thus, for finding QPCs we search for frequency triplets, (f_1, f_2, f_3) for which the bicoherence is high (i.e., close to unity) and the biphase is small (i.e., close to noise level).

In Fig. 2 we show the bicoherence $b_Z(f_1, f_2)$ and the biphase $\Delta\phi_Z$, where Z is the Fourier transform of the GW strain data, in terms of (f_3, f_2) for a stretch of time that was flagged by BCV as a likely candidate for the presence of bilinear or nonlinear coupling noise. Here the bicoherence is strong at $(f_3, f_2) = (495, 10)$ Hz. This indicates that excess noise is being created by a QPC-like nonlinear coupling between $f_2 \approx 10$ Hz, which is known to be a bounce mode of a test mass, and $f_1 = f_2 + f_3 \approx 505$ Hz, which is the violin mode of the same test mass. In Fig. 3 we show plots of the same two quantities on a stretch of data where the bicoherence is strong at $(f_3, f_2) \approx (505.2, 3.8)$ Hz. This is strong evidence for nonlinear coupling noise between $f_2 \approx 3.8$ Hz, which is known to be a mechanical resonance of a test mass, and $f_1 = f_2 + f_3 \approx 509$ Hz, which is the violin mode of that test mass. (Note that different test mass suspensions have somewhat different violin mode frequencies.)

4. Effect of noise artifacts on astrophysical searches

Once a strong indication of bilinear or nonlinear coupling is found as the potential source of a noise artifact, steps must be taken to establish the causal nature of the effect, find its origin and then either remove it or correct its behavior, as the case may be. (More details on this procedure will be reported in a future work, where the improvement in the sensitivity of specific astrophysical searches in aLIGO data brought about by using BCV will be provided.) When it is not possible to completely remove it or correct it steps must be taken to mitigate its effect on the GW channel or on astrophysical searches. A brief example of one such step is presented in this section.

Increased noise, e.g, in the form of side-bands of the violin modes or line harmonics, that is reasonably stationary can hurt the sensitivity of a search for continuous-wave signals from spinning neutron stars [13] or, possibly, a stochastic GW background [14, 15]. On the other hand, transient noise artifacts can adversely affect the sensitivity of astrophysical burst or compact binary coalescence (CBC) searches by increasing their background [10, 22, 23]. For CBC searches, signal discriminatory tests, such as the chi-square test [24], help in keeping the background under control, especially for low-mass searches. For high-mass searches, where the search templates [8] are short in duration and have very few cycles, much like some of the noise transients themselves, these tests are known to be less powerful, and other tests must be devised to supplement them. One such idea was explored in Ref. [25], which found that for a class of these noise artifacts that can be modelled as sine-Gaussian bursts, the time-lag of the CBC templates that get triggered by a burst depends in a (semi-analytically) predictable way on the burst's central frequency and the template's chirp mass. (The chirp mass of a binary with component masses m_1 and m_2 is defined as $(m_1 m_2)^{3/5}/(m_1 + m_2)^{1/5}$.) As shown in Fig. 4, the SNR of a trigger from such a glitch falls off with decreasing template chirp-mass (and increasing time-lag). For a CBC signal, the SNR typically falls much faster with time-lag (see, e.g., Ref. [26]). These characteristics can be capitalized upon by machine-learning algorithms to complement the ability of traditional chi-square tests to discriminate such glitches from real CBC signals [27, 28].

5. Conclusion

Coupling of disturbances at low frequencies, such as the bounce and roll modes, and those at higher frequencies, e.g., the violin modes or power-line harmonics, can adversely affect the detector sensitivity in the band where we expect to find astrophysical signals. Noise of this type can create a stable background or one that ebbs and flows slowly with changing anthropogenic, environmental (e.g., wind-speed variation, microseismic, etc.) or operating conditions, unless it is recognized and removed. Here we demonstrated how BCV can find the presence of such noise couplings and showed how higher order statistics can be used to diagnose certain aspects of those couplings. By way of their power for finding bilinear and nonlinear couplings and diagnosing some of their properties these statistics make useful additions to the set of detector characterization tools that are already being employed for improving aLIGO and AdV data quality. Characterization of interferometer subsystems and environmental disturbances, and reduction of low-level correlated noise, are critical to ensuring good sensitivity to astrophysical searches. With the advent of the first observation run in the advanced detector era, its need has never been more important.

Acknowledgments

We would like to thank Fred Raab for extensive discussions and guidance. We thank T. Isogai, N. Christensen, P. Ajith for collaboration on the BCV code. We also thank T. Dal Canton, P. Fritschel, D. Macleod, T. J. Massinger, L. Nuttal, S. Penn, P. Saulson, R. Schofield, D. Sigg, J. Smith, and S. Whitcomb for helpful discussions. Finally, we thank Jess McIver for discussions

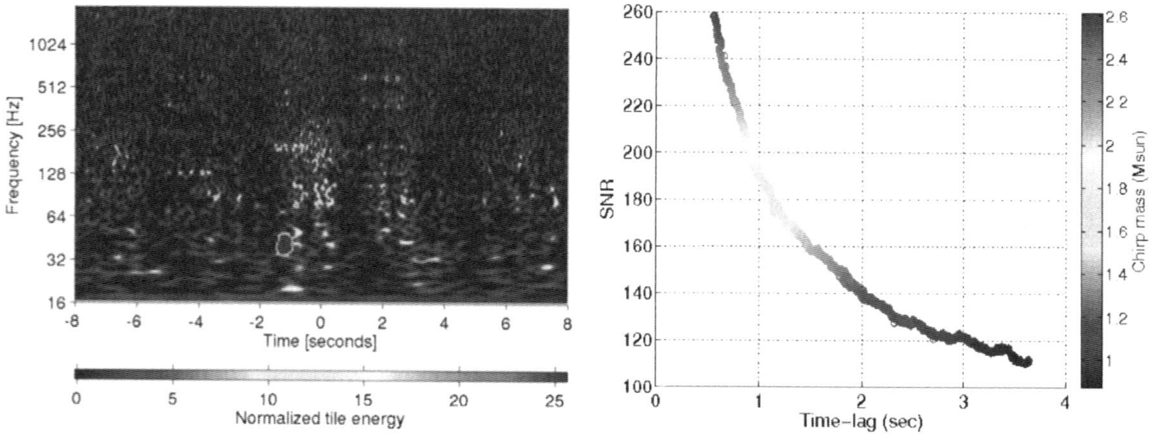

Figure 4. A noise glitch found by Omicron (left figure) and the effect of one such glitch, as modeled by a sine-Gaussian [25] with quality factor, $Q = 10$ and $f_0 = 100$ Hz, on a low-mass CBC template bank (right figure). The time-lag in the latter figure is the time duration between the central time of the glitch and the end-time of the CBC template triggered by that glitch. The signal-to-noise ratio (SNR) of a glitch is maximum when the glitch overlaps with that part of the template where the instantaneous frequency of the latter equals f_0. As the chirp-mass decreases the time-lag increases because the duration of the template after an instantaneous frequency of f_0 increases.

and for carefully reading the manuscript and making several useful and extensive comments on it. This work was supported in part by NSF awards PHY-1206108 and PHY-1506497. The LIGO Document ID of this paper is LIGO-P1500181.

References

[1] J. Aasi *et al.* [LIGO Scientific Collaboration], Class. Quant. Grav. **32**, 074001 (2015) doi:10.1088/0264-9381/32/7/074001 [arXiv:1411.4547 [gr-qc]].
[2] F. Acernese *et al.* [VIRGO Collaboration], Class. Quant. Grav. **32**, no. 2, 024001 (2015) doi:10.1088/0264-9381/32/2/024001 [arXiv:1408.3978 [gr-qc]].
[3] J. Aasi *et al.* [LIGO Scientific and VIRGO Collaborations], arXiv:1304.0670 [gr-qc].
[4] J. Powell, D. Trifir, E. Cuoco, I. S. Heng and M. Cavagli, Class. Quant. Grav. **32**, no. 21, 215012 (2015) doi:10.1088/0264-9381/32/21/215012 [arXiv:1505.01299 [astro-ph.IM]].
[5] J. Abadie *et al.* [LIGO and VIRGO Collaborations], Phys. Rev. D **85**, 082002 (2012) doi:10.1103/PhysRevD.85.082002 [arXiv:1111.7314 [gr-qc]].
[6] J. Aasi *et al.* [LIGO Scientific and VIRGO Collaborations], Phys. Rev. D **89**, no. 10, 102006 (2014) doi:10.1103/PhysRevD.89.102006 [arXiv:1403.5306 [gr-qc]].
[7] D. Talukder, S. Bose, S. Caudill and P. T. Baker, Phys. Rev. D **88**, no. 12, 122002 (2013) doi:10.1103/PhysRevD.88.122002 [arXiv:1310.2341 [gr-qc]].
[8] J. Aasi *et al.* [LIGO Scientific and VIRGO Collaborations], Phys. Rev. D **87**, no. 2, 022002 (2013) doi:10.1103/PhysRevD.87.022002 [arXiv:1209.6533 [gr-qc]].
[9] M. G. Aartsen *et al.* [IceCube and LIGO Scientific and VIRGO Collaborations], Phys. Rev. D **90**, no. 10, 102002 (2014) doi:10.1103/PhysRevD.90.102002 [arXiv:1407.1042 [astro-ph.HE]].
[10] J. Aasi *et al.* [LIGO Scientific and VIRGO Collaborations], Class. Quant. Grav. **32**, no. 11, 115012 (2015) doi:10.1088/0264-9381/32/11/115012 [arXiv:1410.7764 [gr-qc]].
[11] L. Nuttall *et al.*, Class. Quant. Grav. **32**, no. 24, 245005 (2015) doi:10.1088/0264-9381/32/24/245005 [arXiv:1508.07316 [gr-qc]].
[12] J. Aasi *et al.* [VIRGO Collaboration], Class. Quant. Grav. **29**, 155002 (2012) doi:10.1088/0264-9381/29/15/155002 [arXiv:1203.5613 [gr-qc]].
[13] J. Abadie *et al.* [LIGO Scientific and VIRGO Collaborations], Phys. Rev. D **85**, 022001 (2012) doi:10.1103/PhysRevD.85.022001 [arXiv:1110.0208 [gr-qc]].

[14] J. Aasi *et al.* [LIGO Scientific and VIRGO Collaborations], Phys. Rev. D **91**, no. 2, 022003 (2015) [arXiv:1410.6211 [gr-qc]].
[15] D. Talukder, E. Thrane, S. Bose and T. Regimbau, Phys. Rev. D **89**, no. 12, 123008 (2014) doi:10.1103/PhysRevD.89.123008 [arXiv:1404.4025 [gr-qc]].
[16] P. Ajith et al., Phys. Rev. D 89, 122001 (2014).
[17] V. Tiwari *et al.*, Class. Quant. Grav. **32**, no. 16, 165014 (2015) [arXiv:1503.07476 [gr-qc]].
[18] J. M. Mendel, Proceedings of the IEEE, Volume:79, Issue: 3 (1991).
[19] S. Chatterji, L. Blackburn, G. Martin and E. Katsavounidis, Class. Quant. Grav. **21**, S1809 (2004) [gr-qc/0412119].
[20] S. Chatterji, "The search for gravitational wave bursts in data from the second LIGO science run," Ph. D. thesis, Massachusetts Instititute of Technology (2005).
[21] J. W. A. Fackrell, European Signal Processing Conference, 1996. EUSIPCO 1996. 8th.
[22] J. Aasi *et al.* [LIGO Scientific and VIRGO and NINJA-2 Collaborations], Class. Quant. Grav. **31**, 115004 (2014) doi:10.1088/0264-9381/31/11/115004 [arXiv:1401.0939 [gr-qc]].
[23] T. Dayanga and S. Bose, arXiv:1311.4986 [astro-ph.IM].
[24] B. Allen, Phys. Rev. D **71**, 062001 (2005) doi:10.1103/PhysRevD.71.062001 [gr-qc/0405045].
[25] T. Dal Canton et al., Class. Quantum Grav. 31, 015016 (2014).
[26] B. S. Sathyaprakash and S. V. Dhurandhar, Phys. Rev. D **44**, 3819 (1991). doi:10.1103/PhysRevD.44.3819
[27] S. Kapadia, T. Dent, in preparation.
[28] S. Bose, S. Dhurandhar, A. Gupta, A. Lundgren, in preparation.

Journal of Physics: Conference Series **716** (2016) 012008 doi:10.1088/1742-6596/716/1/012008

Comparing interferometry techniques for multi-degree of freedom test mass readout

Katharina-Sophie Isleif, Oliver Gerberding, Moritz Mehmet, Thomas S Schwarze, Gerhard Heinzel, Karsten Danzmann

Leibniz Universität Hannover, Max Planck Institute for gravitational physics (Albert Einstein Institute), Callinstr. 38, 30167 Hannover, Germany

E-mail: katharina-sophie.isleif@aei.mpg.de

Abstract. Laser interferometric readout systems with $1\,\mathrm{pm}/\sqrt{\mathrm{Hz}}$ precision over long time scales have successfully been developed for LISA and LISA Pathfinder. Future gravitational physics experiments, for example in the fields of gravitational wave detection and geodesy, will potentially require similar levels of displacement and tilt readouts of multiple test masses in multiple degrees of freedom. In this article we compare currently available classic interferometry schemes with new techniques using phase modulations and complex readout algorithms. Based on a simple example we show that the new techniques have great potential to simplify interferometric readouts.

High precision interferometry with a dynamic range over multiple interference fringes is the core metrology technique for the space-based low-frequency gravitational wave detector LISA [1]. This technology is also well suited as auxiliary readout system for ground-based detectors, for example for suspension point interferometry [2]. One very promising concept is to apply such a readout not only to a single test mass in one direction, but to all degrees of freedom of multiple test masses in an accelerometer or gradiometer configuration. It is expected that this will make it possible to significantly increase the sensitivity of current accelerometers using electrostatic readout [3]. This would be of great benefit for future geodesy missions that aim to improve on the gravity sensing capabilities of GRACE-FO [4] and GOCE [5].

Different techniques are available and currently under investigation that can be used to construct such readout systems. In the following, we will shortly introduce a number of relevant techniques and discuss their advantages and disadvantages on the basis of conceptual sample implementations. In the end we will briefly compare the techniques and determine the most promising candidate for future research.

Classic heterodyne interferometry already achieves the desired performance with kHz beat notes as used in LISA Pathfinder (LPF) [6], or with MHz beat notes as used in LISA [1, 7] and GRACE-FO [4]. The usage of active stabilisation schemes and several post precessing corrections brings the performance down to $1.42\,\mathrm{pm}/\sqrt{\mathrm{Hz}}$ at 3 mHz [8].

Deep phase modulation (DPM) interferometry is the first example of a technique that can be classified as sinusoidal phase modulation homodyne interferometry. Only one single-frequency laser is split and in one arm a very deep phase modulation ($m \gtrsim 2\pi$) is applied. The DC photodetector signal therefore decreases and several harmonics of the modulation frequency are detected. By performing an IQ-demodulation the complex amplitudes for all harmonics can be recovered. A sophisticated fit algorithm uses this information to extract the four parameters modulation depth and phase, and interferometric amplitude and phase. Considering an electrical null-measurement, using one optical signal split after the photodiode, a phase performance of $1\,\mathrm{pm}/\sqrt{\mathrm{Hz}}$ at 0.1 mHz could be achieved. This is currently limited by white digitisation noise [9, 10]. DPM can be used to simplify the optical set-ups for the laser preparation which are quite complex in comparable kHz heterodyne interferometers, such as in LPF.

We also introduce two multiplexing techniques, **digitally enhanced heterodyne interferometry (DEHeI)** and **digitally enhanced homodyne interferometry (DEHoI)**. DEHeI combines the classical heterodyne interferometer with a digital pseudo-random noise (PRN) high speed phase modulation. It uses the binary phase shift keying (BPSK) technique, which introduces a phase shift by either zero or π on the optical signal in the interferometer path used for the phase measurement of test masses. The beam is then interfered with a frequency shifted local oscillator. It has been shown with DEHeI that signals reflected from various objects with cm-distance and entering the same photodetector can be distinguished from each other due to their different travel time by using high speed PRN phase modulations. The optical complexity decreases but GHz electronics are required for the modulation and demodulation, the detection and the signal digitisation. High speed DEHeI achieved a phase performance of $3\,\mathrm{pm}/\sqrt{\mathrm{Hz}}$ at 10 Hz using an active clock jitter correction [11, 12]. The DEHoI scheme allows for further simplification by removing the second laser frequency. It uses the quadrature phase shift keying (QPSK) scheme to inject two perpendicular PRN phase modulations onto the light. The achieved displacement measurement noise floor is $3\,\mathrm{pm}/\sqrt{\mathrm{Hz}}$ at 4 Hz [13, 14]. Because of the multiplexing capability both, DEHeI and DEHoI, are promising candidates for multichannel interferometry with easily duplicatable optical heads. Furthermore, digital interferometry (DI) is by design insensitive to stray light which makes it very attractive for multi test mass readouts with small dimensions.

Another sinusoidal phase modulation homodyne interferometry technique is **deep frequency modulation (DFM)**. DFM uses only one laser source which is strongly modulated in its frequency. Assuming that the test mass readout consists of an unequal arm length interferometer, the photodetector monitors an optical power signal which is similar to the typical DPM signal, but here the effective phase modulation depth depends on the strength of the frequency modulation and on the arm length mismatch. The DPM fit algorithm can easily be adopted to read out this new kind of interferometers [15–17]. DFM can be used to simplify the set-ups required for the laser preparation delivering kHz beat note signals and it simplifies the optical heads due to intrinsic self-homodyning. First experimental results indicate that pm level performance is feasible [17], but this remains to be verified.

The requirements for a **multi-degree of freedom readout** of various test masses are: The optical head should be as simple and compact as possible with improved sensitivity in comparison to electro-static readout. The phase measurement should be sensitive to differential wavefront sensing (DWS) such that tilts can be measured as well. An insensitivity to stray and scattered light is always advantageous, because spurious reflections spoil the phase measurement. The set-up should be easily duplicatable.

Fig. 1 shows that the complexity of the optical set-ups can be drastically simplified by new interferometer techniques like DI or DFM interferometry, especially because no local reference interferometer is required that measures the phase fluctuations due to the injection fibers or the laser preparation. The amount of phase stable components and photodiodes is reduced and the need for ultra stable fiber couplers can be removed by using self-homodyning techniques as well. DWS can be used if a local oscillator beam is present in the optical head. This requires either a second laser beam or the interference with the laser beam itself. At this point one should remark that a DWS readout requires quadrant photodetectors which are currently not available

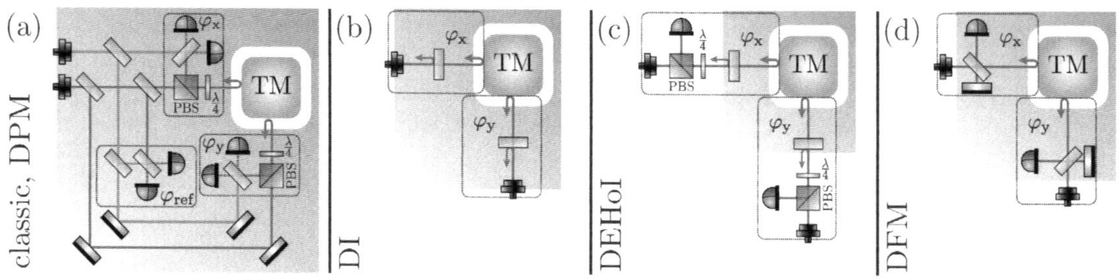

Figure 1. Shown are the optical set-ups for the readout of two longitudinal (and four angular) degrees of freedom of one test mass: Non-multiplexing classic interferometry and also deep phase modulation (DPM) is shown in (a), a possible set-up that can be realised by both digital interferometer (DI) techniques is shown in (b) and two set-ups using the self-homodyning scheme that can be implemented by using digitally enhanced homodyne interferometry (DEHoI) or deep frequency modulation (DFM) interferometry are shown in (c) and (d).

in such small sizes that DEHoI signals with several GHz rates could be detected. Using the alternative DI set-up, the reflections of the test mass and the reference mirrors are interfered with an oscillator beam behind the injection fiber. In this case, multiple signals can be collected by only one photodetector, but DWS cannot be used since the wavefront is destroyed by the fiber coupling. Angular fluctuations could be measured, however, by combining multiple longitudinal degrees of freedom measurements.

Phase modulation interferometer techniques also simplify the set-ups for the laser preparation as shown in Fig. 2. Two acousto-optical modulators (LPF) or two lasers (LISA), can be replaced by one electro-optical modulator and the DPM technique or completely removed by using the DFM technique. This enables a kHz readout without the need for an optical pathlength difference (OPD) stabilisation. However, it remains to be investigated which other stabilisation schemes are required for DPM and DFM. Laser preparation is only needed once, but separate OPD stabilisations are necessary for each pair of fibers feeding light to the optical readout.

Conclusion

Table 1 shows a short comparison of the interferometer techniques and their sensitivities. In terms of optical complexity and excluding GHz electronics and photodiodes the DFM technique might be the most promising candidate for a multi-degree of freedom readout of multiple test

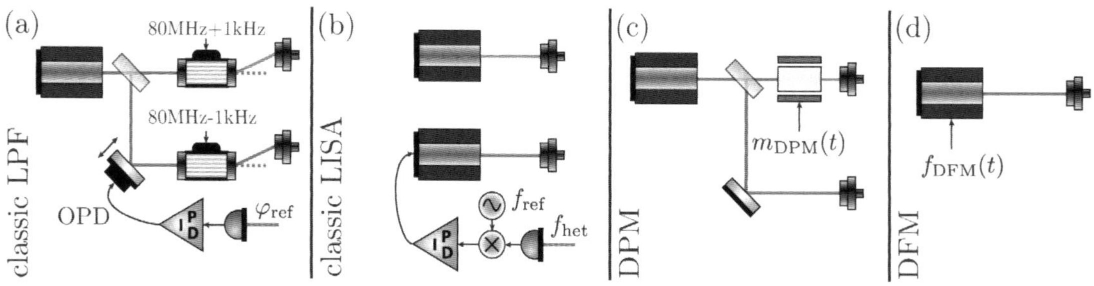

Figure 2. Laser light preparation for various interferometer types. (a) shows the LPF-style modulation for kHz beat notes. The LISA-style laser preparation is used to prepare MHz beat notes and is shown in (b). Inlets (c) and (d) show the application of DFM or DPM to simplify kHz frequency readouts.

masses, even though it does not provide the stray light insensitivity of DI. Further experiments are needed to validate and improve the quoted sensitivities and to determine the set-up complexities of actual implementations. Other potential optical techniques for test mass readout that we have not included here are for example fiber micro-cavity based sensors [18] and optical leaver/shadow sensing [19].

Table 1. Comparison of the new interferometer techniques deep phase modulation (DPM), digital interferometry (DI) and deep frequency modulation (DFM) with the classic heterodyne interferometer. n denotes the number of test masses within a satellite.
[1]based on experimental results, [2]the difference of one optical signal split after photodetector, [3]the difference of one optical signal split before photodetector, *assuming set-ups shown in Fig. 1.

technique	local reference interferometers	electronic bandwidth modulation	readout	DWS capability*	sensitivity [pm/$\sqrt{\text{Hz}}$]
classic	$\geq n$	–	k/MHz	yes	1.4 at 3 mHz[1] [6]
DPM	$\geq n$	kHz	kHz	yes	1.0 at 0.1 mHz[1,2] [10]
DI	0	GHz	GHz	no	3.0 at 10 Hz[1] [12]
DEHoI	0	GHz	homodyne	yes	1.0 at 20 Hz[1] [13]
DFM	0	kHz	kHz	yes	2.5 at 1 Hz[1,3] [17]

Acknowledgments

The authors would like to thank the DFG Sonderforschungsbereich (SFB) 1128 Relativistic Geodesy and Gravimetry with Quantum Sensors (geo-Q) and the International Max-Planck Research School (IMPRS) for financial support.

References

[1] Danzmann K *et al.* 2013 The Gravitational Universe: Whitepaper for the ESA L2/L3 selection
[2] Dahl K, Heinzel G, Willke B, Strain K A, Goßler S and Danzmann K *Classical and Quantum Gravity* 095024
[3] Touboul P, Foulon B and Willemenot E 1999 *Acta Astronautica* **45** 605 – 617
[4] Sheard B S, Heinzel G, Danzmann K, Shaddock D A, Klipstein W M and Folkner W M 2012 *Journal of Geodesy* **86** 1083–1095
[5] Drinkwater M R, Haagmans R, Muzi D, Popescu A, Floberghagen R, Kern M and Fehringer M 2006 *Proceedings of the 3rd international GOCE user workshop* pp 6–8
[6] Heinzel G *et al. Classical and Quantum Gravity* S581
[7] Gerberding O *et al. Review of Scientific Instruments* 074501
[8] Dehne M, Tröbs M, Heinzel G and Danzmann K *Opt. Express* 27273–27287
[9] Heinzel G, Guzmán Cervantes F, García Marin A F, Kullmann J, Feng W and Danzmann K 2010 *Optics express* **18** 19076–19086
[10] Schwarze T S, Gerberding O, Cervantes F G, Heinzel G and Danzmann K 2014 *Opt. Express* **22** 18214–18223
[11] Shaddock D A *Opt. Lett.* 3355–3357
[12] Isleif K S, Gerberding O, Köhlenbeck S, Sutton A, Sheard B, Goßler S, Shaddock D, Heinzel G and Danzmann K 2014 *Opt. Express* **22** 24689–24696
[13] Sutton A J, Gerberding O, Heinzel G and Shaddock D A 2012 *Opt. Express* **20** 22195–22207
[14] de Vine G, Rabeling D S, Slagmolen B J J, Lam T T Y, Chua S, Wuchenich D M, McClelland D E and Shaddock D A 2009 *Opt. Express* **17** 828–837
[15] Kissinger T, Charrett T O and Tatam R P *Opt. Express* 9415–9431
[16] Gerberding O 2015 *Optics Express* **23** 14753–14762
[17] Isleif K S, Gerberding O, Schwarze T S, Mehmet M, Heinzel G and Cervantes F 2016 *Opt. Express* **24** 1676–1684
[18] Smith D, Pratt J and Howard L 2009 *Review of Scientific Instruments* **80** 035105
[19] Carbone L *et al.* 2012 *Classical and Quantum Gravity* **29** 115005

Fast response electromagnetic follow-ups from low latency GW triggers

E J Howell[1], Q Chu[1], A Rowlinson[2], H Gao[3], B Zhang[4], S J Tingay[5,6], M Boër[7] and L Wen[1]

[1]School of Physics, University of Western Australia, Crawley WA 6009, Australia
[2]CSIRO Astronomy and Space Science, Sydney, Australia
[3]Department of Astronomy, Beijing Normal University, Beijing 100875, China
[4]Department of Physics and Astronomy, University of Nevada Las Vegas, NV 89154, USA
[5]International Centre for Radio Astronomy Research, Curtin University, Perth, WA 6845, Australia
[6]ARC Centre of Excellence for All-sky Astrophysics (CAASTRO)
[7]CNRS - ARTEMIS, boulevard de l'Observatoire, CS 34229, 06304 Nice Cedex 04, France

E-mail: E-mail:eric.howell@uwa.edu.au

Abstract. We investigate joint low-latency gravitational wave (GW) detection and prompt electromagnetic (EM) follow-up observations of coalescing binary neutron stars (BNSs). Assuming that BNS mergers are associated with short duration gamma ray bursts (SGRBs), we evaluate if rapid EM follow-ups can capture the prompt emission, early engine activity or reveal any potential by-products such as magnetars or fast radio bursts. To examine the expected performance of extreme low-latency search pipelines, we simulate a population of coalescing BNSs and use these to estimate the detectability and localisation efficiency at different times before merger. Using observational SGRB flux data corrected to the range of the advanced GW interferometric detectors, we determine what EM observations could be achieved from low-frequency radio up to high energy γ-ray. We show that while challenging, breakthrough multi-messenger science is possible through low latency pipelines.

1. Introduction

The first observations by the advanced generation of interferometric gravitational wave (GW) detectors is set to ignite a new era of astronomy in which GW triggered follow-ups by facilities operating outside the GW spectrum will become common place. Complementary observations from electromagnetic (EM), neutrino or high energy particle facilities would not only improve the confidence of a GW detection, but also maximise the science return via a wealth of additional astrophysical information.

The advanced generation of detectors will be led by Advanced LIGO (aLIGO) [1] during the second half of 2015 and followed by Advanced Virgo (AdV) [2] a year later [3]. Coalescing systems of BNSs are one of the most promising sources of GWs in terms of detection [4]; at design sensitivity (around 2018), the expected detection rates of these sources are of order 0.4 to 400 BNS/year [5, 6]. One important aspect of these well-modeled inspiralling systems is that a detection can be made tens of seconds before the merger if enough waveform cycles can be detected to boost the signal-to-noise ratio [7, 8]. This could allow an *extreme low-latency* alert to be sent out to EM facilities as near real-time as possible to catch the fading EM signature. Such

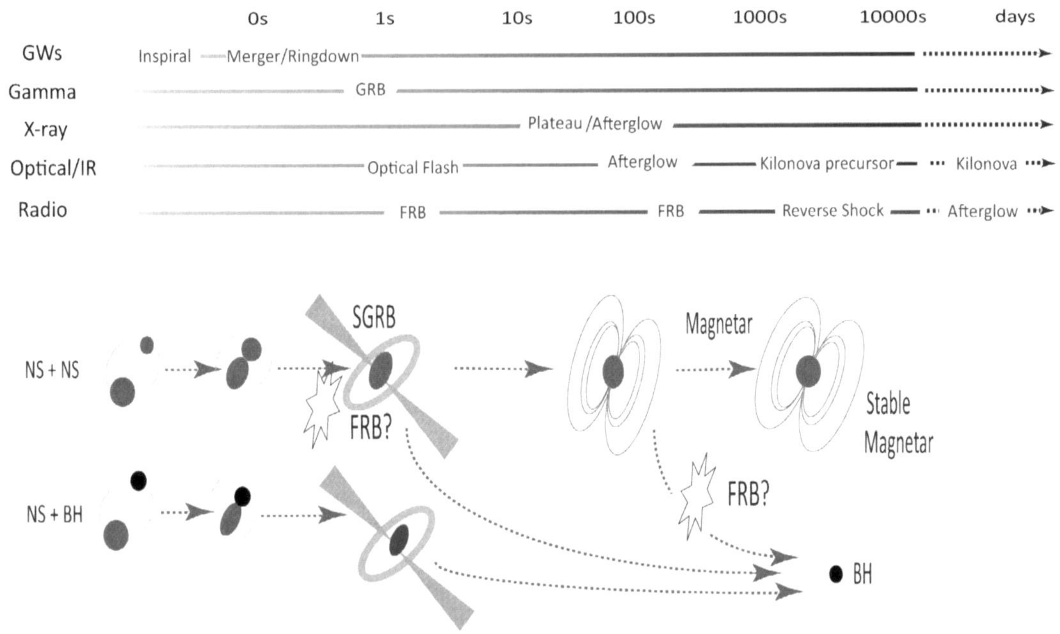

Figure 1. A cartoon illustrating some of the possible EM pathways, timescales and end products of coalescing systems of NSs and BHs. Adapted from Chu et al. [9].

prompt follow-ups could provide valuable insight into the inner workings of these cataclysmic events.

Here we report on the results of a study conducted by Chu et al. [9] which examined the capability of a range of EM instruments for rapid follow-ups from extreme low-latency GW triggers. We present here only the main results assuming an aLIGO/AdV network at design sensitivity. We refer the interested reader to Chu et al. for a complete analysis including results from larger GW Network configurations.

2. The EM pathways for BNS mergers

Figure 1 summarises some of the popular scenarios for the expected EM signatures from BNS mergers. These are based on the proposed link between short duration gamma ray bursts (SGRBs) and the merger of of compact objects [10, 11, 12, 13, 14]. The predictions are highly uncertain and could lead to various pathways and end products. The anticipated onset time, duration, and observable wavelengths are shown in the figure.

Figure 1 shows that SGRBs could be accompanied by an optical flash or a fast radio burst (FRB) [15]; an FRB could also result from the collapse of a merger product to a black hole [16]. If a stable or supra-massive magnetar is formed, the long lived X-ray plateaus observed in many SGRBs could be the signature of direct dissipation of magnetar wind energy [17, 18, 19]. At later times, predictions for an optical/IR kilonova [20] are supported though recent observations [21, 22]. For a relatively small opacity, a kilonova precursor may also appear soon after the merger [23].

SIGNAL TYPE	TARGET	TIME RANGE (s)	Gamma		X-ray	Optical/IR	Radio	
			High (> 10 MeV)	Low (< 10 MeV)			High (> 1 GHz)	Low (< 1 GHz)
Jet Related	SGRB Prompt	0-5	✓	✓	?	?		
	SGRB Afterglow	10s->	✓		✓	✓	✓	?
	Reverse Shock	60-10000	?	?	?	?	✓	?
Isotropic Ejecta Related	Kilonova	1000 ->			?	✓		
	Ejecta Afterglow	10000 ->			?	?	?	?
Other Signals	Magnetar Wind Dissipation	1-1000	?	✓	✓	?	?	?
	FRB (early)	-5 -10					?	?
	FRB (late)	10-10000					?	?

Figure 2. Some of the multimessenger observations that could be achieved for our follow-up scenario. The signals are grouped into jetted, isotropic or other (or unknown) signal types; the approximate emission time ranges around merger are also given. As shown by the legend, **ticks** indicate the EM observations with observational support; **question marks** indicate those that have been proposed through theoretical models; **dark cells** indicate there are EM instruments available in the band with sufficiently fast responses, exposure times and FoVs to cover or tile the GW error box within the emission time-range.

3. Accessing the low-latency GW detector response and EM follow-up

Since the initial LIGO era a number of low-latency GW trigger-generation pipelines have been proposed and tested [24, 25, 8, 26, 27]. The low latency localization pipeline `BAYESTAR` [28, 29] has shown that localization can be achieved in seconds with potential speed ups proportional to the number of CPU cores. For success, wide-field EM telescopes capable of rapid response within tens of seconds and with fields-of-view (FoV) comparable to the error area of GW source sky directions (of order hundreds of $\deg^2$) will be critical. To test the EM-follow up prospects we simulated a total of 200,000 BNS sources, drawing free parameters of the waveform and the sky location of each event from standard distributions. Around 10^3 BNS detections were obtained and the localization accuracy of each was determined using the formalism of [30]. Detailed statistics of the detected sample are provided in Chu et al. [9]. For an aLIGO/AdV network the error region required to contain 50% of the detections, assuming sources are detected 40s before merger, was around 1000 $\deg^2$. We assume this estimate, along with a 40s latency to send out the GW trigger.

To explore prompt EM follow-up observations, we estimate the fluxes at different wavelengths by extrapolating data from observed EM emissions associated with SGRBs and convert to 200 Mpc. We then consider the FoVs, sensitivities and response times for a range of EM telescopes to estimate their ability to capture a prompt EM signature.

4. Results and discussion

Figure 2 shows that at low frequency radio (MWA [31]; FoV:610 $\deg^2$@150 MHz; response in tens of secs) and both high and low-energy γ-ray (e.g. Fermi [32]; GBM FoV:30000 $\deg^2$, LAT FoV:6564 $\deg^2$) the large FoVs and the fast response potential of MWA show that post-prompt (> 5s) breakthrough observations could be realised [33]. In the X-ray, small FoV instruments such as *Swift* are hindered by the large GW error region; a requirement for reconstructed sky locations to overlap with nearby galaxies [34] could improve this situation.

In the optical, even for a wide field, fast response telecope like ZTF ([35]; FoV:47 deg^2; exposure 30s; 20.5-21 mag), to cover the 1000 deg^2 error region with an exposure times of 30s would require hundreds of seconds. Similar instruments will require sophisticated tiling strategies and would benefit from a co-ordinated observational scheme (see [36] for search strategies using the iPTF). A wide FoV X-ray instrument capable of rapid pointing, such as the NASA proposed 900 deg^2 FoV soft X-ray instrument ISS-Lobster [37], would be invaluable for such an approach.

ACKNOWLEDGMENTS

EJH acknowledges support from a UWA Research Fellowship and LW support from the Australian Research Council.

[1] Aasi J *et al.* 2015 *Classical and Quantum Gravity* **32** 074001 (*Preprint* 1411.4547)
[2] Acernese F *et al.* (VIRGO) 2015 *Class.Quant.Grav.* **32** 024001 (*Preprint* 1408.3978)
[3] Aasi J *et al.* 2013 *astro-ph: 1304.0670* (*Preprint* 1304.0670)
[4] Thorne K S 1987 *Three Hundred Years of Gravitation* ed Hawking, S W & Israel, W (Cambridge University Press) pp 330–458
[5] Abadie J *et al.* 2010 *Classical and Quantum Gravity* **27** 173001 (*Preprint* 1003.2480)
[6] Coward D M, Howell E J, Piran T, Stratta G, Branchesi M, Bromberg O, Gendre B, Burman R R and Guetta D 2012 *MNRAS* **425** 2668–2673 (*Preprint* 1202.2179)
[7] Manzotti A and Dietz A 2012 *ArXiv e-prints* (*Preprint* 1202.4031)
[8] Cannon K *et al.* 2012 *ApJ* **748** 136 (*Preprint* 1107.2665)
[9] Chu Q, Howell E, Rowlinson A *et al.* 2015 *submitted to MNRAS.* (*Preprint* 1509.06876)
[10] Paczynski B 1986 *ApJL* **308** L43–L46
[11] Eichler D, Livio M, Piran T and Schramm D N 1989 *Nature* **340** 126–128
[12] Gehrels N, Sarazin C L, O'Brien P T *et al.* 2005 *Nature* **437** 851–854 (*Preprint* arXiv:astro-ph/0505630)
[13] Berger E *et al.* 2005 *Nature* **438** 988–990 (*Preprint* astro-ph/0508115)
[14] Bloom J S *et al.* 2006 *ApJ* **638** 354–368 (*Preprint* astro-ph/0505480)
[15] Totani T 2013 *PASJ* **65** L12 (*Preprint* 1307.4985)
[16] Zhang B 2014 *ApJL* **780** L21 (*Preprint* 1310.4893)
[17] Rowlinson A *et al.* 2010 *MNRAS* **409** 531–540 (*Preprint* 1007.2185)
[18] Rowlinson A, O'Brien P T, Metzger B D, Tanvir N R and Levan A J 2013 *MNRAS* **430** 1061–1087 (*Preprint* 1301.0629)
[19] Zhang B 2013 *ApJL* **763** L22 (*Preprint* 1212.0773)
[20] Metzger B D, Martínez-Pinedo G, Darbha S, Quataert E, Arcones A, Kasen D, Thomas R, Nugent P, Panov I V and Zinner N T 2010 *MNRAS* **406** 2650–2662 (*Preprint* 1001.5029)
[21] Berger E, Fong W and Chornock R 2013 *ApJL* **774** L23 (*Preprint* 1306.3960)
[22] Tanvir N R, Levan A J, Fruchter A S, Hjorth J, Hounsell R A, Wiersema K and Tunnicliffe R L 2013 *Nature* **500** 547–549 (*Preprint* 1306.4971)
[23] Metzger B D, Bauswein A, Goriely S and Kasen D 2015 *MNRAS* **446** 1115–1120 (*Preprint* 1409.0544)
[24] Abadie J *et al.* 2012 *A&A* **541** A155 (*Preprint* 1112.6005)
[25] Buskulic D *et al.* 2010 *Classical and Quantum Gravity* **27** 194013
[26] Luan J, Hooper S, Wen L and Chen Y 2011 *ArXiv e-prints 1108.3174* (*Preprint* 1108.3174)
[27] Hooper S, Chung S K, Luan J, Blair D, Chen Y and Wen L 2012 *PRD* **86** 024012 (*Preprint* 1108.3186)
[28] Singer L and Price L 2015 *in preparation*
[29] Singer L P, Price L R, Farr B *et al.* 2014 *ApJ* **795** 105 (*Preprint* 1404.5623)
[30] Wen L and Chen Y 2010 *Phys. Rev. D* **81** 082001
[31] Tingay S J *et al.* 2013 *PASA* **30** e007 (*Preprint* 1206.6945)
[32] Meegan C *et al.* 2009 *ApJ* **702** 791 URL http://stacks.iop.org/0004-637X/702/i=1/a=791
[33] Howell E J, Rowlinson A, Coward D M *et al.* 2015 *PASA* **32** e046 (*Preprint* 1511.02959)
[34] Evans P A *et al.* 2012 *ApJS* **203** 28 (*Preprint* 1205.1124)
[35] Bellm E 2014 *The Third Hot-wiring the Transient Universe Workshop* ed Wozniak P R, Graham M J, Mahabal A A and Seaman R pp 27–33 (*Preprint* 1410.8185)
[36] Singer L P, Kasliwal M M, Cenko S B *et al.* 2015 *ApJ* **806** 52
[37] Camp J, Barthelmy S, Blackburn L, Carpenter K G, Gehrels N, Kanner J, Marshall F E, Racusin J L and Sakamoto T 2013 *Experimental Astronomy* (*Preprint* 1304.3705)

Journal of Physics: Conference Series **716** (2016) 012010 doi:10.1088/1742-6596/716/1/012010

Experiments driving theory: Gravitational wave detection and the two body problem in general relativity

Bala R. Iyer

International Centre Theoretical Sciences, Tata Institute Fundamental Research, Bangalore, 560012, India

E-mail: bala.iyer@icts.tes.in

Abstract. Over the last two decades, the search for gravitational waves from coalescing compact binaries by detectors like LIGO and Virgo has crucially required and consequently spurred tremendous progress in the two body problem in general relativity. A broad brush overview of these major developments and the current status of these significant results is presented.

1. Introduction

The problem of motion in general relativity (GR) goes back to 1916 to the post Newtonian (PN) works of Einstein, Droste, and De Sitter. Due to difficulties related to the correct treatment of internal structures of bodies only with the 1938 work of Einstein, Infeld and Hoffmann (EIH) did the GR N-body problem reach its first stage of maturity as discussed in the books of Fock, Infeld & Plebanski, Landau & Lifshitz [1]. In the 1916 paper exploring physical implications of GR, Einstein proposed the existence of gravitational waves (GW) as one of its *important* consequences [2]. Soon after, Einstein (1918) calculated the flux of energy far from source; the famous quadrupole formula, and discussed the related radiation reaction or radiation damping [3]. In 1922 Eddington pointed the inapplicability of this derivation for self-gravitating systems [4]. The complication for the self-gravitating case is fundamental: For self-gravitating systems orders in velocity are related to orders in non-linearity. By the virial theorem, $\Phi = GM/R$ is of the same order as v^2. So reaction terms of order $(v/c)^5$ in a linear theory will be accompanied by terms $(v/c)^3\,\Phi/c^2$, $(v/c)^1\,(\Phi/c^2)^2$ for self-gravitating systems. Thus, higher order v/c or PN calculations (v^2/c^2 is designated as 1PN) requires dealing with higher order non-linearities of Einstein's equations (EE). Landau Lifshiftz (1941) and Fock (1955) extended the quadrupole formula to weakly self-gravitating systems and these constitute two different approaches to GW generation today. Chandrasekhar [5] was first to show conceptually that energy and angular momentum radiated as GW was correctly balanced by the loss of mechanical energy and angular momentum. His work gave astrophysicists confidence that GR was physically reasonable and well behaved. But there was a problem. In the gauge he used, some terms at 2PN were divergent. These divergences cast doubt on the validity of Chandrasekhar's treatment for more mathematically demanding relativists.

2. Discovery of the binary pulsar 1913+16 and its implications

In 1974, R.Hulse and J.Taylor discovered the first binary pulsar 1913+16 [6]. Radio pulsar timing observations allow one to reconstruct the orbit and the related inspiral of this system due to emission of GW. This provides high quality data that is proof that GW exist leading to a Nobel Prize in 1993 [7]. The prospects of testing PN theory against Hulse-Taylor system once again revived more critical questions regarding existing treatment of GW. Does it apply to orbital motion of the two neutron stars (NS) even though it does not apply to their internal structure with strong gravitational fields? Even for weak fields is it a valid approximation of GR due to the divergent terms in the PN equations? One needs methods to treat weak orbital fields without assumption on internal fields and Damour showed that orbits and interactions of stars are independent of compactness modulo tidal distortions (*effacement* of internal structure in GR). The high quality binary pulsar data forced a revisit to approximation methods in GR to remedy the mathematical shortcomings in the existing approaches [8]. Discovery of binary systems with strong self gravity mandated improved approaches to the N-body problem: modern versions of EIH. 1913+16 obliged theorists to go beyond 1PN relativistic effects in the equations of motion (EOM); to 2.5PN ie inclusion of terms of $\mathcal{O}(v^5/c^5)$ beyond Newtonian EOM. Insights of a newer generation more comfortable with techniques in field theory to deal with divergences helped. Damour critically looked at the problem and proposed Riesz regularisation to deal with ultra-violet (UV) divergences and iterated EE to sufficient order of non-linearity to obtain EOM of compact binaries including v^5/c^5 terms (1983). Thorne's review [9] summarized state of the art in techniques involved and results then available. Binary pulsars like 1913+16 provide *direct* observational proof that gravity propagates at speed of light and has a quadrupolar structure. They provide accurate tests in the strong field regime of relativistic gravity and prove that GR is valid beyond the quasi-stationary, weak-field regime. It probed for the first time regimes involving radiative effects and strong fields. GR passed all binary pulsar tests with flying colors. The limits on existence of dipole radiation from binary neutron star systems are not strong. However, stringent limits on dipole radiation (test of quadrupolar structure of gravitation radiation damping) exist currently from di-symmetric pulsar- white dwarf systems, the best being from PSR J1738+0333. Binary pulsar population crucially underlie the estimated number of GW events for GW detectors like LIGO and Virgo and the existence of binary neutron star sources emitting GW for hundreds of million years before coalescing spectacularly in the sensitivity bandwidth of these detectors.

3. Prototype Sources: Coalescing Compact Binaries

The late inspiral and merger epochs of compact binaries of neutron stars or black holes provide us possible strong sources of GW for terrestrial Laser Interferometer GW Detectors like LIGO and Virgo in the 'high' frequency range 10 Hz - 10 kHz. We have guaranteed sources for the GW detectors if there are *enough* of them. The waveform is a chirp (amplitude and frequency increasing with time). GW are weak signals buried in the stronger noise of the detector. One requires matched filtering (MF) both for their detection or extraction and parameter estimation or characterisation. Success of MF requires an accurate model of the signal using GR and favours sources like coalescing compact binaries (CCB) made of neutron stars and black holes over sources like supernovae. This led to spectacular theoretical progress in 2-body problem in GR complementing spectacular progress in the GW detection endeavours [10, 11]. When LIGO was funded in early nineties, and efforts to construct accurate CCB waveforms for NS-NS, BH-BH started, it was soon realised that since CCB were highly relativistic in the final stages of inspiral and merger, far higher order PN accurate waveforms would be needed to accurately describe GW at these late epochs than the 2.5PN adequate for binary pulsar work. Physical insights were essential to simplify the goals and achieve the required accuracy for waveforms to construct template banks.. They include: (i) garden variety inspiralling compact binaries (ICB) would have radiated away their eccentricity and be moving in quasi-circular orbits during the late inspiral, (ii) since matched filtering is sensitive to the phase it is more important to first control higher order phasing than higher order amplitudes i.e. use Newtonian amplitude & best available phasing the so called restricted waveform, (iii) the inspiral can be treated in the adiabatic approximation as sequence of circular orbits which allows one to treat separately the *radiation reaction* effects and the *conservative* effects, (iv) one can go to higher PN orders in the inspiral without getting technically bogged down in controlling the much more difficult higher order conservative PN terms, (v) for compact objects the effects of finite size and quadrupole distortion induced by tidal interactions are of order 5PN. Hence, neutron stars and black holes can be modelled as point particles represented by Dirac δ-functions. Thus modelling ICB waveforms in the adiabatic approximation involves three problems: (a) Motion - given a binary system, iterate EE to discuss conservative motion of the system. Compute conserved energy E (and angular momentum (AM) J for the eccentric case); (b) Generation - given the motion of the binary system on a fixed orbit, iterate EE to compute multipoles of the gravitational field and hence the far zone (FZ) flux of energy $\mathcal{L}$ (and FZ flux of AM $\mathcal{J}$) carried by GW; (c) Radiation Reaction (RR) - given the conserved energy (and AM) radiated flux of energy (and AM), *assume* the balance equations to compute the effect of radiation on the orbit. Compute evolution of frequency $F(t)$ and phase $\phi(t)$ (and radial separation $r(t)$). Non-linear evolution of the orbital phase due to higher order gravitational radiation reaction is the crucial ingredient in constructing templates. Higher order phasing is equivalent to inclusion of higher order gravitational radiation reaction (GRR). 3PN (v^6/c^6) flux beyond leading quadrupole determines 3PN (v^6/c^6) RR relative to leading RR at 2.5PN (v^5/c^5). This became a primary driver for research in the two body problem in GR related to higher PN order EOM (effect of conservative acceleration terms), higher PN order of energy and AM fluxes (effect of radiation reaction acceleration terms), very high order fluxes in the test particle limit, very high order computation of the gravitational self force, numerical relativity for the whole inspiral, merger, ringdown (IMR) phases and finally effective one body for its associated analytical description [10, 12].

4. Approximation Schemes, regimes of validity

There are three main methods to deal with two body problem in GR: Post Newtonian (PN) approach, Perturbation theory (including Gravitational Self force (GSF)) and Numerical Relativity (NR). PN approach expands dynamics and waveforms in powers of v/c. It is valid for any mass ratio, in the near-zone (NZ) and in principle for slow motion, which for self-gravitating bodies implies large separations. Implementation of PN approximation is facilitated if its viewed as downstream of the Post - Minkowskian (PM) approximation, Non-linearity expansion or expansion in G. Unlike the PN approximation, the PM approximation is valid all over the weak-field region including infinity. The implementation of PM approximation is facilitated by use of the Multipole (M) expansion or expansion in irreducible representation of the rotation group using symmetric trace free (STF) tensors or equivalently tensor spherical harmonics. Perturbation theory is suitable to describe motion and radiation of a small body moving around a large body. Here one expands EE around a black hole metric rather than Minkowski metric in the ratio of small mass to big mass. This approach can analytically deal with *extreme mass ratio* inspiral (EMRI), & *quasi-normal mode (QNM) ringing.* Going beyond, the back reaction of the gravitational self-field of the small body modifes its geodetic motion leading to corrections to perturbation results and computed by the GSF formalism. Finally, NR solves EE on computer for coalescence, merger and ringdown beyond inspiral. It can be used for any mass ratio, separation or velocity. The range of validity is constrained by computational resources and accuracy requirements on the numerical solutions. Successful wave-generation formalisms are a subtle cocktail of the above different approaches. The most successful Multipolar Post Minkowskian- Post Newtonian (MPM-PN) is employed to deal with *arbitrary mass ratio* inspiral in the slow-motion weak field regime. In this approach, there are two independent aspects addressing two different problems: (i) the general method (MPM expansion) applicable to extended or fluid sources with compact support, based on the mixed PM and multipole expansion (asymptotically) matched to some PN (slowly moving, weakly gravitating, small-retardation) source. Infra-red (IR) divergences arising from the retardation expansion are dealt with by analytic continuation, (ii) the particular application to describe inspiralling compact binaries (ICB) by use of point particle models. Self-field regularisation to deal with ultra-violet (UV) divergences arising from use of Delta functions to model point particles include Riesz, Hadamard partie finie and dimensional regularisation [10].

5. Implication for Ground Based GW detectors

The availibility of the 2PN EOM from binary pulsar work supplemented by a new insight into treatment of cubic non-linearities facilitated the computation of 2PN phasing of ICB by two independent methods: MPM-PN (Blanchet, Damour, Iyer); Direct Integration of Relaxed Einstein equations (DIRE) (Will, Wiseman). The control of 3PN order was more formidable since ambiguities arise in regularisation methods for the self-field (Hadamard regularisation) that worked till 2PN. Only after almost a decade of struggle and by the use of the gauge invariant *dimensional regularisation* was the problem finally resolved and completed. Computation of 3.5PN GW flux requires and includes control of all the non-linear couplings between multipole moments up to order 3.5PN for general matter sources; those couplings involve in addition to many 'instantaneous' terms, the important contributions of tails, tails-of-tails (weakly dependent on past history) and the non-linear memory (strongly dependent on past history)[10, 12]. A new complication arises in the 4PN EOM that has recently been addressed in the ADM approach by Damour, Jaranowski, Schäfer (2014) [13]. At 4PN, there appear irreducible IR divergences (in any gauge) in the calculation of PN expanded 4PN Hamiltonian of binary system related to fundamental breakdown in formal NZ expn of flat spacetime gravitational propagator found by Blanchet and Damour (1988). At 4PN, impossible in any gauge to express NZ metric and 2-body EOM as functional of instantaneous state of the material source [1].

For NS-NS binaries which are inspiral dominated in the sensitive bandwidth of the present detectors, 3PN conservative acceleration and 3PN RR beyond the leading 2.5PN RR description of inspiral is adequate. For BH-BH binaries whose last stable orbit (LSO) lie in the sensitive detector bandwidth the inspiral phase is not adequate. The merger and ringdown need to be crucially modelled as accurately as possible. This requires NR or an extension of the PN description by effective one body or phenomenological approximants. Extensions of the non-spinning results are available for spinning BH binaries (since BH are spinning), quasi-eccentric binaries (there exist astrophysical mechanisms like the Kozai mechanism leading to binaries that could have eccentricity), tidal effects (NS-NS binaries would tidally distort in their final stages), BH horizon flux (though small could be important at higher PN orders). Other approaches include: Direct Integration of Relaxed EE (DIRE) [15] Strong field point particle limit [16] Effective Field Theory (EFT) [17], Gravitational Self Force (GSF) approaches for EMRI's [18] NR-AR and Self force-PN comparisons.

In the test particle limit, analytical results are available to 22PN for energy flux in the Schwarzschild case [19], 11PN in the Kerr case [20] using the Mano, Suzuki, Takasugi functional series formalism. In the Kerr case 20PN energy flux is also know numerically [21]. These are important for parameter estimation in eLISA. Analytical results for black hole horizon absorbed flux are known to 22.5PN in the Schwarzschild case [20, 21]. and 20PN for Kerr [21]. Experimental mathematics, (high accuracy numerics+ PSLQ integer relation algorithm)

[1] Added in Proof: More recently, the 4PN EOM has been investigated by L. Bernard et al [14] using a Fokker action.

Journal of Physics: Conference Series **716** (2016) 012010 doi:10.1088/1742-6596/716/1/012010

can obtain analytic forms of high PN coefficients (involving Euler-Mascheroni gamma, logs of primes, Riemann zeta function at integers) of linear in mass ratio of binding energy [22].

6. Effective one Body approach

Resummation methods like Padé approximants can be used to extend numerical validity of PN expansions (at least) up to the LSO. Effective-one-body (EOB) approach of Buonanno and Damour introduced in 1998 is a particular non-perturbative resummation of PN-expanded EOM to extend validity of PN results beyond the LSO, and up to the merger [1]. Essential inputs are high order PN and perturbation theory results guided by the insight that non-perturbative effects can be captured analytically if key ingredients that enter the two body dynamics and GW emission are properly resummed about the exact test particle results. Each ingredient is crafted through resummation of the PN expansion to incorporate non-perturbative and strong field effects that are lost when dynamics and waveform (WF) are Taylor expanded in PN orders.

Four essential elements of the EOB approach are: (i) *Hamiltonian* $H_{\rm real}$ describing *conservative* part of relative dynamics of 2 BH (ii) *Radiation-reaction force* $\mathcal{F}_\varphi$ describing loss of (mechanical) angular momentum, and energy, of binary system; (iii) Definition of various multipolar components of *"inspiral-plus-plunge" (metric) waveform* $h^{\rm insplunge}_{\ell m}$; (iv) Attachment of subsequent *"Ringdown waveform"* $h^{\rm ringdown}_{\ell m}$ around certain (EOB-determined)"merger time" t_m. EOB provides analytical description of both motion and radiation of CCB from early inspiral, right thro plunge, merger and final ringdown. The complete waveform contrary to concerns is rather simple!! A smooth continuation of inspiral & sharp transition around merger to ringdown. A new improvement involved replacing the standard PN approach of additive WF by by a *multiplicative* WF; factorized resummed waveforms consisting of a physically motivated *product* of Newtonian waveform and various independent relativistic contributions providing parameter a free resummation for multipolar waveform and energy flux [23].

EOB made several bold quantitative and qualitative predictions concerning dynamics of coalescence and the corresponding radiation of GW, (i) a blurred transition from inspiral to a 'plunge' that is just a smooth continuation of the inspiral, (ii) a sharp transition, around the merger of the black holes, between a continued inspiral and a ring-down signal, and (iii) estimates of the radiated energy and of the spin of the final black hole. (iv) larger energy release for spins parallel to the orbital angular momentum, and to a rotation parameter J/E^2 always smaller than unity at the end of the inspiral (so that a Kerr black hole can form right after the inspiral phase). All these predictions have been broadly confirmed by the results of the later numerical simulations performed by several independent groups. The *waveform simplicity* is due to presence of a single characterstic scale close to merger when radiation reaction, orbital and spin precession time scales become of the same order of magnitude, formation of potential barrier around new born BH filtering direct radiation from the merger burst and highly dissipative nature of disturbances in the BH spacetime because of QNM's. EOB with spin and tidal effects are available. EOB allows to combine in a synergetic manner information from other approaches [1].

7. Pretorius and the Numerical Relativity Breakthrough

In 2005 Pretorius [24] produced the first simulation with large number of orbits through merger using modified harmonic coords, compactification of numerical domain at spatial infinity, singularity excision and damping of constraints. With this amazing breakthrough in NR, one has reliable waveforms for the late inspiral and merger parts of the binary evolution which can be used for constructing templates. In eight months other groups using other methods like Baumgarte Shapiro Shibata Nakamura (BSSN) equations and puncture methods followed [25]. These NR waveforms are calibrated and interpreted by the analytical PN inspiral results. There is exciting progress in matching the PN waveforms to the Numerical Relativity ones.

However, BBH simulations are time consuming and NR simulations currently too expensive to solely build required bank of GW templates and densely fill multidimensional space of BBH physical parameters (masses and spins). The state of the art in NR simulation has progressed tremendously over the decade. From short simulations of about 20 orbits; the new one based on Spectral Einstein Code (SpEC) is 25 times longer [26].. First NR simulation of compact binary whose gravitational waveform long enough to cover entire frequency band of Advanced LIGO (125 orbits; Mass ratio 7; Total mass $45.5M_{\odot}$). EOB models uncalibrated or calibrated against smaller NR WF reproduce NR waveform remarkably well while PN inspiral and calibrated IMR phenomenological display greater disagreement.

8. Concluding remarks

At its centenary, GR has not only has passed many stringent tests in flying colors but has consolidated to become an essential tool for describing the universe on manifold scales from satellites and planets to black holes and the big bang. The GW experiment is driving the GR theory. The experimental challenge to detect GW has led to a remarkable progress in the theoretical two body problem in GR. The classic two body problem in GR pushed by the requirements of an unforgiving technology driven fundamental physics experiment has rejuvenated itself to meet the challenges. If *bliss was it in that dawn to be alive* in the last twenty-five years,then, *to be young in the coming decade would be the very heaven!*

8.1. Acknowledgments

I acknowledge support by the AIRBUS Group Corporate Foundation through a visiting professorship, which is part of the Mathematics of Complex Systems chair at ICTS.

References

[1] Damour T 2014 in *Relativity and Gravitation - 100 years after Einstein in Prague* Springer. arXiv:1212.3169
[2] Einstein A 1916 *Sitzungsberichte der Königlich Preussischen Akademie der Wissenschaften Berlin* 688–696
[3] Einstein A 1918 *Sitzungsberichte der Königlich Preussischen Akademie der Wissenschaften Berlin* 154–167
[4] Kennefick D 2007 *Traveling at the speed of thought: Einstein and the quest for gravitational waves* Princeton University Press Princeton.
[5] Chandrasekhar S 1990 *Selected Papers Vol 5* Univ. Of Chicago Press Chicago
[6] Hulse R A, Taylor J H 1975 *Astrophys. J.* **195** L51–L53
[7] Weisberg J M, Nice D J, Taylor J H 2010 *Astrophys. J.* **722** 1030–1034
[8] Damour T 1982 in *Gravitational radiation*, Eds. Deruelle N and Piran T (North Holland, Amsterdam)
[9] Thorne K 1980 Revs. Mod. Phys. **52** 299
[10] Blanchet L 2014 *Living Rev.Rel.* **17** 2
[11] Sasaki M, Tagoshi H 2003 *Living Rev.Rel.* **6** 6
[12] Buonanno A and Sathyaprakash B S 2015 in *General Relativity and Gravitation: A Centennial Perspective*, eds. A. Ashtekar, B. Berger, J. Isenberg and M. A. H. MacCallum (Cambridge University Press)
[13] Damour T, Jaranowski P, Schäfer G 2015 *Phys. Rev.* **D 89** 104455
[14] Bernard L, Blanchet L, Bohe A, Faye G, and Marsat S 2015 arXiv:1512.02876
[15] Will C M and Wiseman A G 1996 *Phys.Rev. D* **54** 4813
[16] Futamase T, Itoh Y 2007 *Living Rev.Rel.*, **10** 2
[17] Rothstein I 2014 *Gen. Rel. Grav.* **46** 1726
[18] Poisson E, Pound A and Vega I 2011 *Living Rev.Rel.* **14** 7
[19] Fujita R 2012 *Prog. Theor. Phys.* **128** 971-992
[20] Fujita R 2015 *Prog. Theor. Exp. Phys.* Issue 3 033E01
[21] Shah A 2014 *Phys. Rev.* **D 90** 044025
[22] Johnson-McDaniel N K, Shah A G and Whiting B F 2015 *Phys. Rev.* **D 92** 044007
[23] Damour T, Iyer B R and Nagar A 2009 *Phys. Rev.* **D 79** 064004
[24] Pretorius F 2005 *Phys. Rev. Lett.* **95** 121101
[25] Choptuik M W, Lehner L, Pretorius F in 2015 *General Relativity and Gravitation: A Centennial Perspective*, eds. A. Ashtekar, B. Berger, J. Isenberg and M. A. H. MacCallum (Cambridge University Press)
[26] Szilagyi B, et al 2015 *Phys. Rev.* **D 91** 124021

Test environments for the GRACE follow-on laser ranging interferometer

A Görth[1], J Sanjuan[2], M Gohlke[2], S Rasch[2], K Abich[2], C Braxmaier[2,3] and G Heinzel[1]

[1] Max Planck Institute for Gravitational Physics, Callinstr. 38, 30169 Hanover, Germany
[2] Deutsches Zentrum fur Luft- und Raumfahrt, Robert-Hooke-Str. 7, 28359 Bremen, Germany
[3] Center of Applied Space technology and Microgravity (ZARM), University of Bremen, Am Fallturm 1, 28359 Bremen, Germany

E-mail: alexander.goerth@aei.mpg.de

Abstract. In the year 2017 a follow-on mission to the very successful joint German/US mission GRACE (Gravity Recovery And Climate Experiment) will be launched. As of this day the two GRACE satellites have successfully been mapping the spatial and temporal varitations of the gravitational field of the Earth by satellite-to-satellite tracking for over a decade. The main science instrument on GRACE and its successor mission GRACE Follow-On which is used to measure the inter-satellite distance changes is a microwave link sensor. However, an additional instrument, the laser ranging interferometer (LRI), will be implemented into the architecture of the GRACE Follow-On satellites as a technology demonstrator. In this paper we will give a brief overview of a fiber-based test environment which is currently used during the assembly, integration and test of the LRI flight hardware.

1. Introduction

The Gravity Recovery And Climate Experiment (GRACE) has been a highly successful satellite mission. Launched in 2002, it has been mapping the gravitational field of the Earth for over a decade and has provided the science community with unprecedented insights into the spatial and temporal variations of the mass distribution of the Earth. The main contributor to these variations is water - as well in liquid as in solid form. Amongst the multitude of published results that could be obtained from GRACE data are publications on the Greenland ice sheet mass loss (e.g. [1], [2]), the global sea level rise which can be associated with the melting of ice caps and glaciers (e.g. [3]) or the depletion of groundwater in regions that notoriously suffer from a shortage of water like the California central valley ([4], [5]) or northern India ([6]). On GRACE a microwave ranging instrument has been used to determine a change in the inter-satellite distance. From this, together with global positioning system (GPS) and accelerometer data, one can derive the gravitational field below the two spacecraft (SC). A complete update of the Earth's gravity field is released about once per month. However, the tandem SC are way past their design mission time of five years and are suffering from battery fatigue. Hence, updates of the gravitational field were increasingly released on an irregular basis.

Anticipating the imminent end of the original GRACE mission the same US/German team has decided to launch a follow-on mission as soon as possible to keep the data gap to a minimum. In August 2017 the GRACE Follow-On mission is planned to be launched. The tight schedule that is caused by this early launch date could only be fulfilled by flying an almost exact copy of the original GRACE SC. While a microwave instrument will remain the main science instrument that is used to determine the inter-satellite distance, a laser ranging interferometer (LRI) will be integrated into the SC as a technology demonstrator. The accuracay of the LRI is anticipated to be $80\,\mathrm{nm}/\sqrt{\mathrm{Hz}}$, which is more than one order of magnitude better than the accuracy of the microwave instrument [7].

Being able to test and calibrate this complex device on ground is crucial during its assembly, integration and test activities. While also subsystems require special test environments, the most sophisticated tests of the LRI can be performed when the two SC are enabled to exchange laser beams. This is referred to as *constellation level* tests. These tests require a test environment which experimentally simulates the complete laser link between the SC independently of their physical orientation in the laboratory. That is why a fiber-based set-up was designed that fulfils exactly these requirements. In the manner of space flight jargon it is referred to as fiber-based *optical ground support equipment* - or short Fiber OGSE. In the following section an overview of the functionality of the Fiber OGSE and a brief introduction into its architecture is given.

2. Description of the Fiber OGSE set-up

The task of the Fiber OGSE is to experimentally simulate the laser link between the two GRACE Follow-On SC. This includes beam properties (see e.g. [7]) as well as static and dynamic SC misalignments. Due to the SC separation of approximately 230 km the beam that is received by one satellite (RX beam) has a nearly spherical wavefront. Since only a small fraction of it eventually reaches the optical bench, the RX beam is well approximated by a top-hat beam. The maximum RX beam power is of the order of a few hundred picowatts and drops as a function of the misalignment between the transmitted beam (TX beam) and the line of sight (LOS):

$$P_{\mathrm{RX}} = 2\frac{P_0}{\omega_{\mathrm{RX}}} r_{\mathrm{ap}}^2 e^{-2\left((\alpha-\alpha_{\mathrm{SC}})^2+(\beta-\beta_{\mathrm{SC}})^2\right)\ell^2/\omega_{\mathrm{RX}}^2}. \tag{1}$$

Here $P_0 \approx 20\,\mathrm{mW}$ is the TX beam power, $\omega_{\mathrm{RX}} \approx 35\,\mathrm{m}$ is the beam radius at the receiving SC, $r_{\mathrm{ap}} = 4\,\mathrm{mm}$ is the radius of the receiving aperture on the OBs and $\ell \approx 230\,\mathrm{km}$ is the separation of the satellites [8]. α_{SC} and β_{SC} describe the mispointing i.e., the angle between the TX beam and the LOS. The actual ranging signal is a frequency modulation of the laser beam, with power P_{RX}, due to Doppler shifts and is of the order of $\pm 4\,\mathrm{MHz}$ [7].

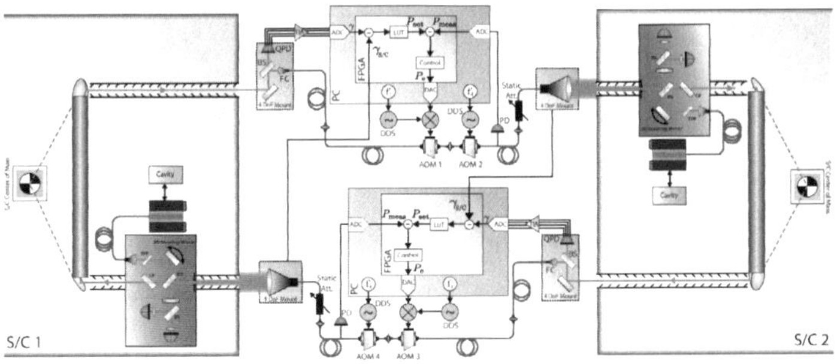

Figure 1. Schematic of the Fiber OGSE as it is mounted between the two (simplified) GRACE Follow-On SC [9].

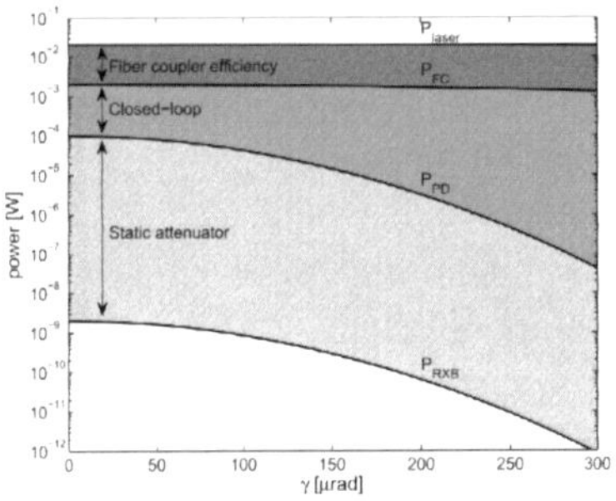

Figure 2. The different attenuation stages of the DPA [9].

Figure 1 shows a schematic of the Fiber OGSE as it will be used between two SC. On the left and right hand side of the figure simplified sketches of the SC with the main parts of the LRI are shown. The upper and lower units of the Fiber OGSE are identical and represent the dual link between the SC. In the upper link the transmitted beam (TX beam) leaves the SC on the top left and is immediately coupled into a fiber. This part of the Fiber OGSE is referred to as the *transmitted beam assembly*. At the top right the light exits the Fiber OGSE through a fiber collimator which generates a laser beam with approximately 2 cm waist radius. The collimator with its tip/tilt and translation mount is called *received beam assembly*. In between these two units is the *dynamic power attenuation unit* (DPA). This contains two acousto-optical modulators (AOMs) and an FPGA-based control loop which controls the AOMs in such a way that the RX power matches the far-field power that the SC on the right side of the picture would detect if both satellites were separated by approx. 230 km (see eq. 1 and [9]). Hereby the AOMs serve as actuators and the power of the radio-frequency (RF) signal that controls them is the controlled variable.

2.1. The Received Beam Assembly

The RX beam assembly consists of a fiber collimator with a lens of focal length $f = 200\,\mathrm{mm}$ in a four degree of freedom mount. The beam that is generated by this collimator has a waist radius of 18 mm ($1/e^2$). The corresponding wavefront curvature was measured with a Hartmann-Shack-sensor and deviates by less than $\lambda/15$, where $\lambda = 1064\,\mathrm{nm}$ from a flat wavefront within the receiving aperture. The power drop in this area is about 10 %. The motorized rotation mount allows for a tilt in yaw and pitch of the collimator of $\pm 26\,\mathrm{mrad}$. This is more than sufficient to simulate static misalignments of the SC as well as SC attitude jitter. To properly align the unit with respect to the SC a two degree of freedom linear stage is also used. A readout mechanism that is based on calibrated potentiometers which are attached to the motors of the rotation mount is used to monitor the angular motion of the unit [9]. When SC jitter is simulated, this information is forwarded to the DPA which takes it into account for the calculation of the AOM gains (see eq. 1).

2.2. The Transmitted Beam Assembly

The TX beam assembly serves two purposes: launch the TX beam into the fiber setup and measure the TX beam angle. The latter is being done with a DC measurement of the TX beam centroid on a 10 mm diameter quadrant photodiode (QPD). The QPD currents are transformed to voltages by a transimpedance amplifier (TIA). These are then fed to analog-to-digital converters (ADC) and then to the FPGA (see fig. 1). Here the voltages of the individual QPD segments are combined to yield the position of the TX beam centroid on the QPD. By applying calibration factors that denote the lever arm of the TX beam the horizontal and vertical angles of the TX beam are obtained. The TX beam assembly is also mounted to a four degree of freedom rotation/translation stage for alignment of the unit and to simulate SC misalignments.

2.3. Dynamic Power Attenuator

The core of the Fiber OGSE is the DPA. A closed-loop control with approximately 50 kHz bandwidth that is implemented in an FPGA is used to generate the correct power level at the receiving SC [9]. For this the FPGA evaluates the TX beam angle and its predefined set-point to eventually apply an amplitude modulated RF signal to one of the AOMs. When Doppler shifts are applied the other AOM is operated in a feed-forward control to keep it at constant gain (with respect to the optical power) and can therefore be neglected for the control-loop. The dynamic attenuation of the laser power is a multi-step process (see fig. 2). The first angle-dependent gain is introduced by the fiber collimator of the TX beam assembly. In order to keep this nonlinear gain to a minimum, a collimator was chosen whose coupling efficiency is rather insensitive to a tilt of the laser beam. The control-loop in the DPA then dynamically attenuates the laser power to the microwatt level and a static attenuator further reduces the signal power to the picowatt range. With this scheme the control-loop can be operated above the electronic noise floor with signal-to-noise ratios of 10 – 10.000 in an angular range of $0 \leq \gamma \leq 225\,\mu$rad [9]. This implies that SC jitter, which is a varying misalignment of the SC, could be simulated with an amplitude of up to $225\,\mu$rad by shifting the set-point of the DPA control-loop. The static part of the SC misalignment (which is in a range of a few mrad [14]) is introduced by deliberately misaligning the RX and TX beam assemblies by using their motorized rotation mounts. However, larger amplitudes of the SC jitter could also be introduced by synchronously varying the tilt of the RX and the TX assemblies with respect to the SC. Because the repeatibility of the used rotation stages would not suffice for this, only the home-made read-out mechanism of the rotation mounts renders this possible.

3. Conclusion

In this paper we gave an overview of the design of the Fiber OGSE which is a test environment that is currently being used during AIT of the GRACE Follow-On LRI. It realistically simulates the laser link between the two SC and allows for the introduction of static and dynamic SC tilts. Due to its flexibility, the setup can also be used to study future mission concepts. The performance of the Fiber OGSE was already proven and the results published [9]. This reference may also be considered by the reader for a more detailled description of the Fiber OGSE. Recently, tests of the LRI initial link acquisition procedure were performed using the Fiber OGSE with mock-up OBs. While this was done before in a similar way ([14],[15]), up to now these tests have been the most realistic of their kind. Especially the accurate imitation of the far-field power as well as the generation of a flat wavefront with homogeneous intensity profile of the RX beam is unprecedented. The results of these tests will be published soon in another paper.

Acknowledgments

The authors would like to thank the Bundesministerium für Bildung und Forschung (BMBF, project number 03F0654B) and the International Max Planck Research School on Gravitational Waves (IMPRS-GW) for partly funding this work.

References

[1] Svendsen P L, Andersen O B and Nielsen A A 2013 *Earth Planet. Sci. Lett.* **264** 24–29
[2] Velicogna I 2009 *Geophys. Res. Lett.* **36** L19503
[3] Chen J L, Wilson C R and Tapley B D 2013 *Nat. Geosci.* **6** 549–52
[4] Famiglietti J S, Lo M, Ho S L, Bethune J, Anderson K J, Syed T H, C S S, de Linage C R and Rodell M 2011 *Geophys. Res. Lett.* **38**
[5] Borsa A A, Agnew D C and Cayan D R 2014 *Science* **345** 1587–1590
[6] Tiwari V M, Wahr J and Swenson S 2009 *Geophys. Res. Lett.* **36**
[7] Sheard B S, Heinzel G, Danzmann K, Shaddock D A, Klipstein W M and Folkner W M 2012 *J. Geod.* **86** 1083–95
[8] Schütze D, Stede G, Müller V, Gerberding O, Bandikova T, Shear B, Heinzel G and Danzmann K 2014 *Opt. Exp.* **22** 24117–32
[9] Sanjuan J, Gohlke M, Rasch S, Abich K, Görth A, Heinzel G and Braxmaier C 2015 *Appl. Opt.* **54** 6682–89
[10] Ward, R et al 2014 *Class. Quantum Grav.* **31**
[11] Schütze D, Farrant D, Shaddock D A, Sheard B S, Heinzel G and Danzmann K 2014 *Rev. Sci. Instrum.* **85**
[12] Schütze D, Müller V, Stede G, Sheard B S, Heinzel G, Danzmann K, Sutton A and Shaddock D A 2014 *Opt. Express* **22** 9324–33
[13] Schütze D 2015 *Intersatellite Laser Interferometry: Test Environments for GRACE Follow-On* Ph.D. thesis Leibniz University Hanover
[14] Wuchenich D, Mahrdt C, Sheard B S, Francis S, Spero R, Miller J, Mow-Lowry C, Ward R, Klipstein W M, Heinzel G, Danzmann K, McClelland D and Shaddock D A 2014 *Opt. Exp.* **22** 11351–66
[15] Wuchenich D M R 2014 *Inter-Satellite Laser Interferometry* Ph.D. thesis Australian National University

Inspiral waveforms for spinning compact binaries in a new precessing convention

Anuradha Gupta[1] **and Achamveedu Gopakumar**[2]
[1]Inter University Centre for Astronomy and Astrophysics, Ganeshkhind, Pune 411007, India
[2]Department of Astronomy and Astrophysics, Tata Institute of Fundamental Research, Mumbai 400005, India

E-mail: anuradha@iucaa.in

Abstract. It is customary to use a precessing convention, based on Newtonian orbital angular momentum $\boldsymbol{L}_{\rm N}$, to model inspiral gravitational waves from generic spinning compact binaries. A key feature of such a precessing convention is its ability to remove all spin precession induced modulations from the orbital phase evolution. However, this convention usually employs a post-Newtonian (PN) accurate precessional equation, appropriate for the PN accurate orbital angular momentum $\boldsymbol{L}$, to evolve the $\boldsymbol{L}_{\rm N}$-based precessing source frame. This motivated us to develop inspiral waveforms for spinning compact binaries in a precessing convention that explicitly use $\boldsymbol{L}$ to describe the binary orbits. Our approach introduces certain additional 3PN order terms in the orbital phase and frequency evolution equations with respect to the usual $\boldsymbol{L}_{\rm N}$-based implementation of the precessing convention. The implications of these additional terms are explored by computing the match between inspiral waveforms that employ $\boldsymbol{L}$ and $\boldsymbol{L}_{\rm N}$-based precessing conventions. We found that the match estimates are smaller than the optimal value, namely 0.97, for a non-negligible fraction of unequal mass spinning compact binaries.

1. Introduction

Inspiralling compact binaries containing spinning black holes (BHs) are plausible sources for the network of second generation gravitational wave (GW) detectors like the advanced LIGO (aLIGO), advanced Virgo, KAGRA, GEO-HF and the planned LIGO-India [1]. The inspiral dynamics and associated GWs from compact binaries can be accurately described using the post-Newtonian (PN) approximation to general relativity [2]. Moreover, an optimal detection technique of *matched filtering* is employed to detect and characterize inspiral GWs from such binaries. In this technique, one cross correlates the interferometric output data with a bank of templates that theoretically model inspiral GWs from spinning binaries. The construction of these templates involves modeling the two GW polarization states, $h_\times(t)$ and $h_+(t)$, associated with such events, in an accurate and efficient manner. At present, GW frequency and associated phase evolution, crucial inputs to compute $h_{\times,+}(t)$, are known to 3.5PN order for non-spinning compact binaries [3] whereas the amplitudes are available to 3PN order [4]. In the case of spinning components, the spin effects enter the dynamics and GW emission via spin-orbit (SO) and spin-spin (SS) interactions [5]. Additionally, $\boldsymbol{S}_1$, $\boldsymbol{S}_2$ and $\boldsymbol{L}$, the two spin and orbital angular momenta, for generic spinning compact binaries precess around the total angular momentum $\boldsymbol{J} = \boldsymbol{L} + \boldsymbol{S}_1 + \boldsymbol{S}_2$ due to SO and SS interactions. This forces substantial modulations of the emitted GWs from inspiralling generic spinning compact binaries [6, 7]. Therefore, it is important to incorporate various spin effects while constructing inspiral GW templates for spinning compact binaries. At present, GW frequency evolution and amplitudes of $h_{\times,+}(t)$ for maximally spinning BH binaries are fully determined to 2.5PN and 2PN orders, respectively, while incorporating all the relevant spin induced effects [6, 8].

There exist inspiral waveforms for precessing binaries, implemented in the LALSIMULATION package of LIGO Scientific Collaboration (LSC) [9], that employ the *precessing convention* of [10]. An attractive feature of this convention is its ability to remove all the spin precession induced modulations from the orbital phase evolution. This allows one to express the orbital phase $\Phi_p(t)$ as an integral of the orbital frequency $\omega(t)$, namely $\Phi_p(t) = \int \omega(t)\, dt$. Therefore, in this convention, the inspiral waveform from precessing binaries can be written as the product of a non-precessing carrier waveform and a modulation term that contains all the precessional effects. The convention involves a *precessing source frame* ($\boldsymbol{e}^l_1, \boldsymbol{e}^l_2, \boldsymbol{e}^l_3 \equiv \hat{\boldsymbol{L}}_{\rm N}$) whose basis vectors satisfy the evolution equations $\dot{\boldsymbol{e}}^l_{1,2,3} = \boldsymbol{\Omega}^l_e \times \boldsymbol{e}^l_{1,2,3}$. The angular frequency Ω^l_e is constructed in such a manner that the three basis vectors $\boldsymbol{e}^l_1, \boldsymbol{e}^l_2$ and $\hat{\boldsymbol{L}}_{\rm N}$ always form an orthonormal triad. Subsequently, $\boldsymbol{\Omega}^l_e$ has to be $\boldsymbol{\Omega}^l_e = \boldsymbol{\Omega}_k - (\boldsymbol{\Omega}_k \cdot \hat{\boldsymbol{L}}_{\rm N})\hat{\boldsymbol{L}}_{\rm N}$, where Ω_k is the usually employed precessional frequency for $\boldsymbol{L}_{\rm N}$. The relevant expression for $\boldsymbol{\Omega}_k$ can be obtained by collecting the terms that multiply $\hat{\boldsymbol{L}}_{\rm N}$ in equation (9) in [10]. This triad defines an orbital phase $\Phi_p(t)$ such that $\boldsymbol{n} = \cos\Phi_p\, \boldsymbol{e}^l_1 + \sin\Phi_p\, \boldsymbol{e}^l_2$, where $\boldsymbol{n}$ is the unit vector along binary separation vector $\boldsymbol{r}$. Furthermore, one can express $\dot{\boldsymbol{n}}$ in the co-moving frame $(\boldsymbol{n}, \boldsymbol{\lambda} = \hat{\boldsymbol{L}}_{\rm N} \times \boldsymbol{n}, \hat{\boldsymbol{L}}_{\rm N})$ as $\dot{\boldsymbol{n}} = \dot{\Phi}_p \boldsymbol{\lambda} + \boldsymbol{\Omega}^l_e \times \boldsymbol{n}$. It was argued in [10] that $\boldsymbol{\Omega}^l_e$ should only be proportional to $\boldsymbol{n}$, leading to $\dot{\boldsymbol{n}} = \dot{\Phi}_p \boldsymbol{\lambda}$. Thus, the adiabatic condition for the sequence of circular orbits, namely $\dot{\boldsymbol{n}} \cdot \boldsymbol{\lambda} = \omega$, gives the desired result, i.e., $\dot{\Phi}_p = \omega$. It should be obvious that the above adiabatic condition can also imply $\dot{\boldsymbol{n}} \cdot \dot{\boldsymbol{n}} = \omega^2$.

In practice, the precessional equation for $\boldsymbol{L}$ is employed to construct $\boldsymbol{\Omega}^l_e$ and to evolve $\boldsymbol{L}_{\rm N}$. As a consequence, $\boldsymbol{\Omega}^l_e$ is no longer proportional to $\boldsymbol{n}$ [11] and this leads to PN corrections to $\dot{\Phi}_p = \omega$ (see section 2.1 of [12] for detailed calculation). These observations motivated us to provide a set of PN accurate equations to obtain temporally evolving quadrupolar order $h_{\times,+}$ for generic spinning compact binaries in an $\boldsymbol{L}$-based precessing convention. In the next section, we present our $\boldsymbol{L}$-based precessing convention and explore its data analysis implications in the later section.

2. Inspiral waveforms via an L-based precessing convention

In this section, we introduce a $\boldsymbol{k}$-based precessing source frame ($\boldsymbol{e}_1$,$\boldsymbol{e}_2$, $\boldsymbol{e}_3 \equiv \boldsymbol{k}$), to develop a $\boldsymbol{k}$-based precessing convention, where $\boldsymbol{k}$ is unit vector along $\boldsymbol{L}$. The precessional dynamics of $\boldsymbol{e}_1, \boldsymbol{e}_2$ and $\boldsymbol{e}_3$ are provided by $\dot{\boldsymbol{e}}_{1,2,3} = \boldsymbol{\Omega}_e \times \boldsymbol{e}_{1,2,3}$, where $\boldsymbol{\Omega}_e \equiv \boldsymbol{\Omega}_k - (\boldsymbol{\Omega}_k \cdot \boldsymbol{k})\, \boldsymbol{k}$ and Ω_k is the usual precessional frequency of $\boldsymbol{k}$. It should be obvious that $\dot{\boldsymbol{e}}_3 = \boldsymbol{\Omega}_e \times \boldsymbol{e}_3$ is identical to $\dot{\boldsymbol{e}}_3 = \boldsymbol{\Omega}_k \times \boldsymbol{e}_3$ as $\boldsymbol{e}_3 \equiv \boldsymbol{k}$. It is possible to construct a $\boldsymbol{k}$-based co-moving triad $(\boldsymbol{n}, \boldsymbol{\xi} = \boldsymbol{k} \times \boldsymbol{n}, \boldsymbol{k})$ and define an orbital phase Φ such that $\boldsymbol{n} = \cos\Phi\, \boldsymbol{e}_1 + \sin\Phi\, \boldsymbol{e}_2$ and $\boldsymbol{\xi} = -\sin\Phi\, \boldsymbol{e}_1 + \cos\Phi\, \boldsymbol{e}_2$. Also, the time derivatives of $\boldsymbol{n}$ is given by $\dot{\boldsymbol{n}} = \dot{\Phi}\, \boldsymbol{\xi} + \boldsymbol{\Omega}_e \times \boldsymbol{n}$. Consequently, the frame independent adiabatic condition for circular orbits, namely $\dot{\boldsymbol{n}} \cdot \dot{\boldsymbol{n}} \equiv \omega^2$, leads to $\omega^2 = \dot{\Phi}^2 + \Omega^2_{e\xi}$, where $\Omega_{e\xi} = \boldsymbol{\Omega}_e \cdot \boldsymbol{\xi}$. This results in the following 3PN accurate differential equation for Φ,

$$\dot{\Phi} = \frac{c^3}{G m}\, x^{3/2} \left\{ 1 - \frac{x^3}{2} \left[\delta_1\, q\, \chi_1 \left(\boldsymbol{s}_1 \cdot \boldsymbol{\xi}\right) + \frac{\delta_2}{q}\, \chi_2 \left(\boldsymbol{s}_2 \cdot \boldsymbol{\xi}\right) \right]^2 \right\}, \tag{1}$$

where $q = m_1/m_2$, $\delta_{1,2} = \eta/2 + 3\,(1 \mp \sqrt{1-4\eta})/4$, $\eta = m_1 m_2/m^2$ and $m = m_1 + m_2$. The PN expansion parameter x is defined as $(G\, m\, \omega/c^3)^{2/3}$. The Kerr parameters χ_1 and χ_2 of the two compact objects of mass m_1 and m_2 specify their spin angular momenta by $\boldsymbol{S}_{1,2} = G m^2_{1,2}\, \chi_{1,2}\, \boldsymbol{s}_{1,2}/c$, where $\boldsymbol{s}_1$ and $\boldsymbol{s}_2$ are the unit vectors along $\boldsymbol{S}_1$ and $\boldsymbol{S}_2$. The use of $\boldsymbol{L}$ to describe binary orbits also modifies the evolution equation for ω (or x). This is because the SO interactions are usually incorporated in terms of $\boldsymbol{s}_1 \cdot \boldsymbol{l}$ and $\boldsymbol{s}_2 \cdot \boldsymbol{l}$, in the literature [6, 10].

These terms require modifications due to the 1.5PN order relation between $\boldsymbol{l}$ and $\boldsymbol{k}$ which can be obtained from equation (8) in [12]. The PN accurate expression for $\dot{x}$ along with the 3PN additional terms is given by equation (9) in [12]. Notice that these additional terms are, for example, with respect to equation (3.16) in [13] that provides PN accurate expression for $\dot{x}$ while invoking $\boldsymbol{l}$ to describe binary orbits.

We now model inspiral GWs from spinning binaries in our $\boldsymbol{k}$-based precessing convention. The expressions for quadrupolar order $h_\times$ and h_+, written in the frame-less convention [6], read

$$h_\times|_{\rm Q}(t) = 2\,\frac{G\,m\,\eta\,x}{c^2\,R'}\,(2\,\xi_x\,\xi_y - 2\,n_x\,n_y)\,, \tag{2a}$$

$$h_+|_{\rm Q}(t) = 2\,\frac{G\,m\,\eta\,x}{c^2\,R'}\,(\xi_x^2 - \xi_y^2 - n_x^2 + n_y^2)\,, \tag{2b}$$

where $\xi_{x,y}$ and $n_{x,y}$ are the x and y components of $\boldsymbol{\xi}$ and $\boldsymbol{n}$ in an inertial frame associated with $\boldsymbol{N}$, the unit vector that points from the source to the detector, while R' is the distance to the binary. These x and y components of $\boldsymbol{\xi}$ and $\boldsymbol{n}$ can be expressed in terms of the Cartesian components of $\boldsymbol{e}_1$ and $\boldsymbol{e}_2$. In order to obtain $h_\times|_{\rm Q}(t)$ and $h_+|_{\rm Q}(t)$, we require to solve numerically the differential equations for $\Phi, x, \boldsymbol{e}_1$ and $\boldsymbol{e}_2$. We use equation (1) for Φ while the differential equation for x, $\boldsymbol{e}_1$ and $\boldsymbol{e}_2$ are given by equations (9) and (13) in [12]. It easy to see that the evolution of $\boldsymbol{e}_1$ and $\boldsymbol{e}_2$ depends upon the time variation of $\boldsymbol{s}_1$, $\boldsymbol{s}_2$ and $\boldsymbol{k}$. Therefore, we also need to solve differential equations for $\boldsymbol{s}_1$, $\boldsymbol{s}_2$ and $\boldsymbol{k}$. These differential equations that include the leading order SO and SS interactions can be obtained from equation (15) in [12].

In practice, we numerically solve the differential equations for $\boldsymbol{e}_1, \boldsymbol{k}$, $\boldsymbol{s}_1$, $\boldsymbol{s}_2$, Φ and x to obtain temporally evolving Cartesian components of $\boldsymbol{\xi}$ and $\boldsymbol{n}$. Note that we do not solve the differential equation for $\boldsymbol{e}_2$. This is because the temporal evolution of $\boldsymbol{e}_2$ can be estimated using the relation $\boldsymbol{e}_2(t) = \boldsymbol{k}(t) \times \boldsymbol{e}_1(t)$. The required initial values for the Cartesian components of $\boldsymbol{e}_1$, $\boldsymbol{k}$, $\boldsymbol{s}_1$ and $\boldsymbol{s}_2$ are given by freely choosing the following five angles: θ_{10}, ϕ_{10}, θ_{20}, ϕ_{20} and ι_0. The initial Cartesian components of $\boldsymbol{s}_1$, $\boldsymbol{s}_2$, $\boldsymbol{k}$ and $\boldsymbol{e}_1$ as functions of the above angles are given by equations (16) in [12]. Note that this choice of initial conditions is influenced by the LALSUITE SpinTaylorT4 code of LSC. Additionally, we let the initial x value to be $x_0 = (G\,m\,\omega_0/c^3)^{2/3}$ where $\omega_0 = 10\pi$ Hz (relevant for aLIGO) and the initial phase Φ_0 to be zero. In what follows, we explore the data analysis implications of these inspiral waveforms that employ the $\boldsymbol{L}$-based precessing convention.

3. Implications of inspiral waveforms in $\boldsymbol{L}$-based precessing convention

We employ the *match* [14] to compare inspiral waveforms constructed via the $\boldsymbol{l}$ and $\boldsymbol{k}$-based precessing conventions. Our comparison is influenced (and justified) by the fact that the precessing source frames of these two conventions are functionally identical. This should be evident from the use of *the same* precessional frequency, appropriate for $\boldsymbol{k}$, to obtain PN accurate expressions for both the $\boldsymbol{l}$-based $\boldsymbol{\Omega}_e^l$ and $\boldsymbol{k}$-based $\boldsymbol{\Omega}_e^l$. Therefore, the *match* estimates probe influences of the additional 3PN order terms present in the differential equations for Φ and x in our approach. Note that these 3PN order terms are not present in the usual implementation of the precessing convention as provided by the LALSUITE SpinTaylorT4 code.

Our match $\mathcal{M}(h_l, h_k)$ computations involve h_l and h_k, the two families of inspiral waveforms arising from the $\boldsymbol{l}$ and $\boldsymbol{k}$-based precessing conventions. The h_l inspiral waveform families are adapted from the LALSUITE SpinTaylorT4 code of LSC while h_k families arise from our approach (equations (2)). We employ the quadrupolar order expressions for $h_{\times,+}$ while computing h_l and h_k in the present analysis. Moreover, the two families are characterized by identical values of m, η, χ_1 and χ_2. Also, the initial orientations of the two spins in the $\boldsymbol{N}$-based inertial frame are also chosen to be identical. The computation of $\boldsymbol{N}\cdot\boldsymbol{l}$ from $\boldsymbol{N}\cdot\boldsymbol{k}$ with the help of equation (8) in

[12] ensures that $\boldsymbol{l}$ and $\boldsymbol{k}$ orientations at the initial epoch are physically equivalent. Therefore, our match computations indeed compare two waveform families with physically equivalent orbital and spin configurations at the initial epoch. Note that we terminate h_l and h_k inspiral waveform families when their respective x parameters reach 0.1 ($r \sim 10\,G\,m/c^2$).

Figure 1 represents the result of our $\mathcal{M}$ computations. The binary configurations have initial dominant SO misalignments $\tilde{\theta}_1(x_0)$ ($\cos^{-1}(\boldsymbol{k} \cdot \boldsymbol{s}_1)$) as 30° and we let the initial orbital plane orientation in the $\boldsymbol{N}$-based inertial frame to take two values leading to edge-on ($\iota_0 = 90°$) and face-on ($\iota_0 = 0°$) binary orientations. For these two configurations, we need to choose θ_{10} to be 30° and 60°, respectively. Moreover, we choose $\phi_{10} = 0°$, $\theta_{20} = 20°$, $\phi_{20} = 90°$. Let us note that $\boldsymbol{l}$ orientations (from $\boldsymbol{N}$) for these configurations will be slightly different from 0° or 90° due to the 1.5PN accurate relation between $\boldsymbol{l}$ and $\boldsymbol{k}$.

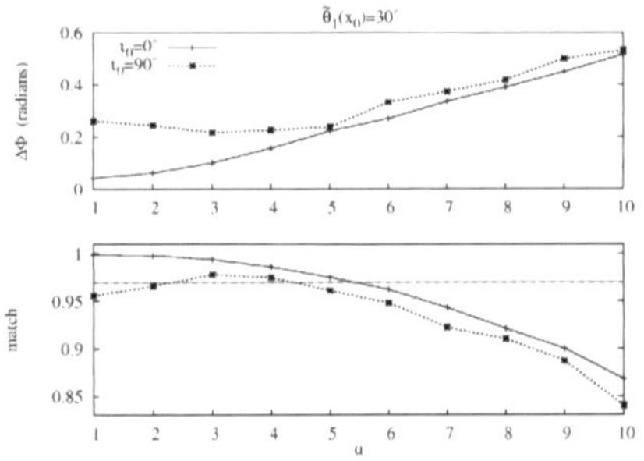

Figure 1. Plots for the accumulated orbital phase ($\Delta\Phi$) and the associated match ($\mathcal{M}$) estimates as functions of the mass ratio q for maximally spinning $m = 30M_\odot$ BH binaries inspiralling in the $[x_0, 0.1]$ frequency interval.

We also plot $\Delta\Phi$, the accumulated orbital phase differences in the frequency interval $[x_0, 0.1]$. We find that the variations in $\mathcal{M}$ estimates are quite independent of the initial orbital plane orientations. We see a gradual decrease in $\mathcal{M}$ values as we increase the q value and this variation is reflected in the gradual increase of $\Delta\Phi$. Incidentally, this pattern is also observed for configurations having somewhat smaller initial dominant SO misalignments. However, the $\mathcal{M}$ estimates are close to unity for tiny $\tilde{\theta}_1(x_0)$ values and this is expected as precessional effects are minimal for such binaries. Therefore, the effect of the above discussed additional 3PN order terms are more pronounced for high mass ratio compact binaries having *moderate* dominant SO misalignments.

We find that the match estimates are less than the optimal 0.97 value for a non-negligible fraction of unequal mass spinning compact binaries. It may be recalled that such an optimal match value roughly corresponds to a 10% loss in the ideal event rate. We, therefore, conclude that the additional 3PN order terms in frequency and phase evolution equations in our approach should not be neglected for a substantial fraction of unequal mass binaries.

References

[1] Sathyaprakash B S and Schutz B F 2009 *Living Rev. Relativity* **12**
[2] Blanchet L 2006 *Living Rev. Relativity* **9** 4
[3] Blanchet L, Damour T, Esposito-Farese G and Iyer B R 2004 *Phys. Rev. Lett.* **93** 091101
[4] Blanchet L, Faye G, Iyer B R and Sinha S, 2008 *Class. Quantum Grav.* **25** 165003
[5] Barker B and O'Connell R 1975 *Phys. Rev. D* **12** 329
[6] Kidder L 1995 *Phys. Rev. D* **52** 821
[7] Apostolatos T A, Cutler C, Sussman G J and Thorne K 1994 *Phys. Rev. D* **49** 6274
[8] Buonanno A, Faye G and Hinderer T 2013 *Phys. Rev. D* **87** 044009; references therein
[9] http://www.ligo.org/index.php
[10] Buonanno A, Chen Y and Vallisneri M 2003 *Phys. Rev. D* **67** 104025
[11] Gupta A and Gopakumar A 2014 *Class. Quantum Grav.* **31** 065014, arXiv:1308.1315
[12] Gupta A and Gopakumar A 2015 *Class. Quantum Grav.* **32** 175002, arXiv:1507.00406.
[13] Bohe A, Marsat S and Blanchet L 2013 *Class. Quantum Grav.* **30** 135009
[14] Damour T, Iyer B R and Sathyaprakash B S 1998 *Phys. Rev. D* **57** 885

Search method for an emission line of a GW background

Atsushi Nishizawa[1*] **and Naoki Seto**[2]

[1] Theoretical Astrophysics 350-17, California Institute of Technology, Pasadena, California 91125, USA

[2] Department of Physics, Kyoto University, Kyoto 606-8502, Japan

E-mail: *anishi@caltech.edu

Abstract. A sharp emission line of a gravitational-wave background (GWB) would be an interesting observational target. Here we study an efficient method to detect a line GWB by correlating data of multiple GW detectors. We find that the width of frequency bin in the data analysis is a critical parameter, and, with the commonly-used value 0.25 Hz, the signal-to-noise ratio could be decreased by up to a factor of 6.6, compared with a finer width of 0.02Hz. By reanalyzing the existing data with a smaller bin width, we might detect a precious line signal from the early universe.

1. Introduction

With the advent of the second-generation GW interferometric detectors, the first detection of a GW would be achieved in a couple of years. At present, the primordial GWs have not been detected yet. However, in the future, the high penetrating power of GWs could become quite advantageous to probe the early universe such as inflation, the subsequent reheating era, high energy physics, and possibly gravity theory beyond general relativity, and extend our observational reach before the big bang nucleosynthesis [1].

To maximize the scientific outputs from the accumulated data of detectors, we should deliberate on methods of data analysis and thoroughly search for stochastic GWBs, not only for given theoretical predictions but also in model independent manners. One of such GWs would be a sharp emission line of a cosmological GWB. The line GWB is not vital for the contemporary standard model of cosmology but is still allowed to exist. To create a line, there needs a sharply defined physical scale such as periodicity, boundaries, or discreteness. These could be realized in the cosmological scenarios with extra dimensions whose topologies are torus or warped between branes [2, 3], or in the massive gravity theories that gravitons have tiny mass [4].

This article is a short summary of our previous work [5], in which we have studied how GW emission lines are constrained by GW observations in a model-independent way. Throughout this paper, we use the unit $c = 1$ for the speed of light.

2. Doppler broadening and GW signals

We assume that a GWB is isotropic in the CMB rest frame but nearly monochromatic, having a sharp line spectrum. Seen from a detector moving relative to the CMB frame with velocity $\vec{v}(t)$, the GWB looks anisotropic because the observed frequency is Doppler-shifted, depending on the propagation direction of a GW $\hat{\Omega}$. The relative velocity $\vec{v}_s$ of the Solar-System barycenter to the CMB rest frame is toward the Galactic coordinate $(\ell, b) = (263.99° \pm 0.14°, 48.26° \pm 0.03°)$ with the magnitude $v_s = (1.230 \pm 0.003) \times 10^{-3}$ $(= 369.1\,\mathrm{km\,s^{-1}})$ [6]. Below, for simplicity, we put $\vec{v}(t) = \vec{v}_s$, neglecting corrections (at most $\leq 10\%$) due to the velocity of the detector relative to the SSB. We also neglect minor relativistic effects and drop the terms of $O(|\vec{v}(t)|^2)$ for the Doppler effect. Then we have GW frequency in the observer's frame

$$f(t, \hat{\Omega}) = f_r [1 - v_s \cos\theta] \ , \tag{1}$$

with $\cos\theta = \hat{\Omega} \cdot \vec{v}_s / v_s$. Therefore, for a given observational frequency, the GW signals come from a ring on the sky (with a fixed θ). The total width of the Doppler broadening is given by $\delta f_D = 2 f_r v_s \sim 0.2\,(f_r/100\,\mathrm{Hz})$ Hz .

We start with the correlation analysis for an anisotropic GWB (not specific to a line GWB) [7]. In the frequency space, the magnitude of a GWB is characterized conventionally by the energy density of GWs per logarithmic frequency bin per steradian divided by the critical energy density of the Universe, $\Omega_{gw}(f; \hat{\Omega})$. The correlation analysis is an efficient method to distinguish a GWB signal from detector noises (see *e.g.* [8–10]). With this method, we prepare data streams of two widely separated detectors with independent noises, and take cross correlations of their Fourier modes. After binning specified by the label k, its central frequency f_k, and its width δf_b $(> T^{-1}$, T: observation time, but much smaller than the all characteristic variation of signals), the expectation value of a signal in each bin k is given by [10]

$$\mu_k = \frac{3H_0^2}{8\pi^2} \frac{T \delta f_b}{f_k^3} Z(f_k). \tag{2}$$

Here we use the Hubble parameter $H_0 = 70\,\mathrm{km\,s^{-1}\,Mpc^{-1}}$ and defined the integral

$$Z(f_k) \equiv \int \frac{d\hat{\Omega}}{4\pi} \Omega_{gw}(f_k; \hat{\Omega}) \gamma(f_k, \hat{\Omega}) \ , \tag{3}$$

and the direction-dependent overlap-reduction function

$$\gamma(f, \hat{\Omega}) \equiv \sum_A F_I^A(\hat{\Omega}) F_J^A(\hat{\Omega}) \exp\left[2\pi i f \hat{\Omega} \cdot \vec{d}_{IJ}\right] \ , \tag{4}$$

where $\vec{d}_{IJ}(t) \equiv \vec{X}_I(t) - \vec{X}_J(t)$ is the distance vector between the Ith detector at the position $\vec{X}_I(t)$ and the Jth detector at $\vec{X}_J(t)$. F_I^A is the antenna pattern function of the Ith interferometer. So far the formulation is generic and we have not assumed any specific spectrum of a GWB.

Now we apply the above expressions specifically to the line GWB at a frequency f_r defined in the CMB frame. From Eq.(1), we can write the observed spectrum by

$$\Omega_{gw}(f; \theta) = \tilde{e}_{gw} f_r \ \delta[f - f_r(1 - v_s \cos\theta)], \tag{5}$$

where $\tilde{e}_{gw}$ is the total energy of the line normalized by the critical density. For a line GWB, we define the overlap-reduction function $\Gamma(f, u)$ that is obtained by integrating $\gamma(f, \hat{\Omega})$ across the ring directions at $u \equiv \cos\theta =$ constant as

$$\Gamma(f, u) \equiv \int_0^{2\pi} \frac{d\phi}{2\pi} \gamma(f, \hat{\Omega}) \ . \tag{6}$$

From Eqs.(2)-(6), replacing the frequencies of slowly-varying functions with f_r and putting $\Gamma(f_k) \equiv \Gamma(f_r, u(f_k))$, we obtain

$$\mu_k \approx \frac{3H_0^2}{16\pi^2} \frac{T \tilde{e}_{gw}}{f_r^3 v_s} \ \delta f_b \ \Gamma(f_k) \ . \tag{7}$$

Note that the frequency f and the sky angle θ (equivalently u) are related by Eq. (1).

This correlation signal has three interesting behaviors (see [5] for more details). Firstly, $\Gamma(f)$ is in general a complex number and has a finite imaginary part. For an isotropic GWB (*e.g.* without a line component), the imaginary part of the correlated signal exactly cancels out due to the directional symmetry of GWs. In contrast, for a line GWB, the imaginary part does not cancel out because of the anisotropy induced by the Doppler effect. Secondly, at high frequency regime, the correlation signal is rapidly oscillating through the phase factor, $\exp\left[2\pi i f\hat{\Omega}\cdot\vec{d}_{IJ}\right]$ in $\Gamma(f_k)$, which strongly depends on the propagation direction $\hat{\Omega}$. The condition $f_{\rm r}(\hat{\Omega}\cdot\vec{d}_{IJ})\sim 1$ and Eq. (1) lead to the characteristic frequency interval δf_c for the oscillation, $\delta f_c \sim v_{\rm s}/d_{IJ} \simeq 0.1\,(3000\,{\rm km}/d_{IJ})$ Hz. Comparing this result with the Doppler width, the number of oscillations in δf_D becomes larger for a higher frequency $f_{\rm r}$. For another detector pair more separated than aLIGOs (d_{IJ} ~3000 km), the characteristic interval δf_c becomes smaller. Thirdly, the frequency dependence of $\Gamma(f)$ changes in time due to the spin and revolution of the Earth because the relative direction from the GWB rest frame (the CMB frame) changes. These motions bring complications into the data analysis. However, they would be useful to distinguish true signals from artificial noise lines produced by instruments.

In Eq.(2), we assumed to take a small bin width δf_b so that the frequency dependence of relevant functions can be neglected within each bin. But this is not always the case. Indeed, the commonly used value by LIGO and VIRGO in [11] is $\delta f_b = 0.25$ Hz, though the characteristic frequency interval is $\delta f_c \sim 0.1$ Hz. For the case with $\delta f_b > \delta f_c$, the correlation signal $|\mu_k|$ averaged out by the integration of the wavy structure in each frequency bin. Then Eq.(7) should be replaced with the following integral

$$\mu_k = \frac{3H_0^2}{16\pi^2}\frac{T\tilde{e}_{\rm gw}}{f_r^3 v_{\rm s}}\int_{\delta f_b} df\,\Gamma(f)\,. \tag{8}$$

3. Sensitivity to a line GWB

To calculate the signal-to-noise ratio (SNR) of the line detection, the noises of the two detectors are assumed to be independent and to have magnitudes much larger than those of the GW signals. Since a detector noise in the correlation signal is slowly varying function in frequency, we can use the same noise formula in [10] for both cases with smaller and larger bin than δf_c. Using Eqs. (7) and (8), the total squared SNR is evaluated as [5]

$$\rho^2 = \frac{9H_0^4}{32\pi^4}\frac{T\tilde{e}_{\rm gw}^2\delta f_b}{f_r^6 v_s^2 P_I(f_r)P_J(f_r)}\sum_k |\Gamma(f_k)|^2 \qquad \text{for } \delta f_b \ll \delta f_c\,, \tag{9}$$

$$\rho^2 = \frac{9H_0^4}{32\pi^4}\frac{T\tilde{e}_{\rm gw}^2\delta f_b}{f_r^6 v_s^2 P_I(f_r)P_J(f_r)}\sum_k \frac{1}{(\delta f_{b(k)})^2}\left|\int_{\delta f_{b(k)}} df\,\Gamma(f)\right|^2 \qquad \text{for } \delta f_b \gg \delta f_c\,. \tag{10}$$

with the one-sided noise spectra $P_I(f)$. For a total observational duration T much longer than a day, the factors involving with $\Gamma(f_k)$ oscillate diurnally. So the factors should be regard as time averaged one.

In Fig. 1, the sensitivity to the normalized energy density $\tilde{e}_{\rm gw}$ is shown as a function of the line frequency $f_{\rm r}$. We assumed a 1 yr observation with two aLIGO detectors at the detection threshold of $\rho = 10$. For their noise spectra $P_{I,J}(f)$, we use the fitting formula given in [12]. For our demonstration here, we selected the following four widths; 0 Hz, 0.1 Hz, 0.25 Hz, and 2.5 Hz. The first one means a sufficiently small width (still satisfying $\delta f_b > T^{-1}$). Figure 1 shows that the sensitivity decreases significantly for larger bin widths due to the cancellation of the wavy structure. The degradation factors from $\delta_b = 0$ Hz case, say, at $f_{\rm r} = 70$ Hz, is 6.8 for

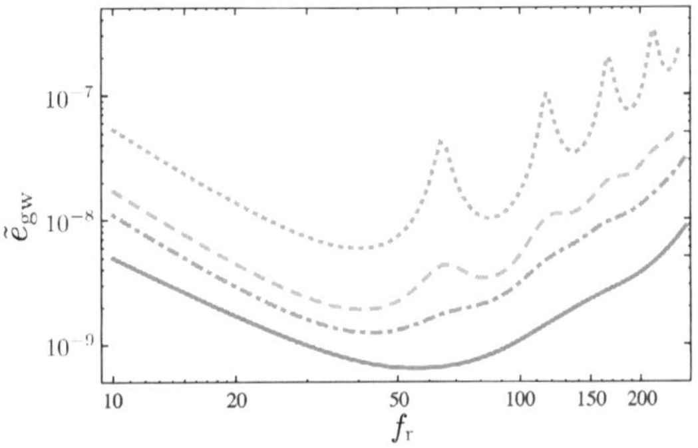

Figure 1. Sensitivity to the normalized energy density $\tilde{e}_{\rm gw}$ of a line GWB, assuming 1 yr observation with two aLIGOs at detection threshold $\rho = 10$. The blue (solid) curve is for the bin width $\delta_b = 0$ Hz. The other curves from the top are with the bin width 2.5 Hz (green, dotted), 0.25 Hz (orange, dashed), 0.1 Hz (red, dotted-dashed), respectively. Adopted from [5].

$\delta f_b = 0.25$ Hz and 22 for 2.5 Hz. We also examined bin widths smaller than δf_c =0.1 Hz. For a line GWB with 10Hz $< f_{\rm r} <$ 600Hz, the degradations (compared with $\delta f_b = 0$) are within 1.1 for $\delta f_b = 0.01$ Hz, 1.2 for 0.02 Hz, 1.5 for 0.04 Hz, 3.7 for 0.1 Hz and 7.9 for 0.25Hz. Therefore, to realize a high SNR, we should set the bin width sufficiently smaller than the characteristic interval $\delta f_c \sim 0.1$ Hz.

4. Conclusions

We have studied the detector signals from a line GWB and suggested the correlation analysis method for efficiently searching for such GW signals. Since the overlap reduction function has fine wavy structures, the correlation signal could be canceled out unless the bin width δf_b is much smaller than the characteristic frequency interval $\delta f_c \sim 0.1$ Hz. Nevertheless, in the standard correlation analysis of GWBs, the commonly used width is 0.25 Hz. By reanalyzing the existing data with a smaller width, we might actually uncover an important signal from the early universe.

Acknowledgments

A. N. is supported by JSPS Postdoctoral Fellowships for Research Abroad (No. 25-180). N. S. is supported by JSPS (No. 24540269) and MEXT (No. 24103006).

References

[1] Maggiore M 2000 *Phys.Rept.* **331** 283–367 (*Preprint* gr-qc/9909001)
[2] Chodos A and Detweiler S 1980 *Phys. Rev. D* **21** 2167–2170
[3] Seahra S S, Clarkson C and Maartens R 2005 *Physical Review Letters* **94** 121302 (*Preprint* gr-qc/0408032)
[4] Gümrükçüoğlu A E, Kuroyanagi S, Lin C, Mukohyama S and Tanahashi N 2012 *Classical and Quantum Gravity* **29** 235026
[5] Nishizawa A and Seto N 2015 *Phys. Rev. D* **91** 122001 (*Preprint* 1503.00354)
[6] Aghanim N *et al.* (Planck Collaboration) 2013 (*Preprint* 1303.5087)
[7] Allen B and Ottewill A C 1997 *Phys.Rev.* **D56** 545–563 (*Preprint* gr-qc/9607068)
[8] Christensen N 1992 *Phys.Rev.* **D46** 5250–5266
[9] Flanagan E E 1993 *Phys.Rev.* **D48** 2389–2407 (*Preprint* astro-ph/9305029)
[10] Allen B and Romano J D 1999 *Phys.Rev.* **D59** 102001 (*Preprint* gr-qc/9710117)
[11] Aasi J *et al.* (LIGO Scientific Collaboration, VIRGO Collaboration) 2014 *Phys.Rev.Lett.* **113** 231101 (*Preprint* 1406.4556)
[12] Sathyaprakash B and Schutz B 2009 *Living Rev.Rel.* **12** 2 (*Preprint* 0903.0338)

Single-Source Gravitational Wave Limits From the J1713+0747 24-hr Global Campaign

T. Dolch[1,2] (for the NANOGrav Collaboration), J. A. Ellis[3], S. Chatterjee[2], J. M. Cordes[2], M. T. Lam[2], C. Bassa[4,5], B. Bhattacharyya[5], D. J. Champion[6], I. Cognard[7], K. Crowter[8], P. B. Demorest[9], J. W. T. Hessels[4,10], G. Janssen[4], F. A. Jenet[11], G. Jones[12], C. Jordan[5], R. Karuppusamy[6], M. Keith[5], V. I. Kondratiev[4,13], M. Kramer[6,14], P. Lazarus[6], T. J. W. Lazio[3], D. R. Lorimer[15], D. R. Madison[2,16], M. A. McLaughlin[15], N. Palliyaguru[15], D. Perrodin[17], S. M. Ransom[16], J. Roy[5,18], R. M. Shannon[19], R. Smits[4], I. H. Stairs[8], B. W. Stappers[5], D. R. Stinebring[20], K. Stovall[21], J. P. W. Verbiest[6,22], W. W. Zhu[6]

[1]Department of Physics, Hillsdale College, 33 E. College Street, Hillsdale, MI 49242, USA
[2]Astronomy Department, Cornell University, Ithaca, NY 14853, USA
[3]Jet Propulsion Laboratory, California Institute of Technology, 4800 Oak Grove Drive, Pasadena, CA 91106, USA
[4]ASTRON, the Netherlands Institute for Radio Astronomy, Postbus 2, 7990 AA, Dwingeloo, The Netherlands
[5]Jodrell Bank Centre for Astrophysics, School of Physics and Astronomy, The University of Manchester, Manchester M13 9PL, UK
[6]Max-Planck-Institut für Radioastronomie, Auf dem Hügel 69, D-53121 Bonn, Germany
[7]Laboratoire de Physique et Chimie de l'Environnement et de l'Espace, LPC2E UMR 6115 CNRS, F-45071 Orléans Cedex 02, and Station de radioastronomie de Nançay, Observatoire de Paris, CNRS/INSU, F-18330 Nançay, France
[8]Department of Physics and Astronomy, University of British Columbia, 6224 Agricultural Road, Vancouver, BC V6T 1Z1, Canada
[9]National Radio Astronomy Observatory, 1003 Lopezville Rd., Socorro, NM 87801, USA
[10]Anton Pannekoek Institute for Astronomy, University of Amsterdam, Science Park 904, 1098 XH Amsterdam, The Netherlands
[11]Center for Advanced Radio Astronomy, University of Texas, Rio Grande Valley, Brownsville, TX 78520, USA
[12]Columbia Astrophysics Laboratory, Columbia University, NY 10027, USA
[13]Astro Space Center of the Lebedev Physical Institute, Profsoyuznaya str. 84/32, Moscow 117997, Russia
[14]University of Manchester, Jodrell Bank Observatory, Macclesfield, Cheshire, SK11 9DL, UK
[15]Department of Physics and Astronomy, West Virginia Univ., Morgantown, WV 26506, USA
[16]National Radio Astronomy Observatory, 520 Edgemont Road, Charlottesville, VA 22901, USA
[17]INAF-Osservatorio Astronomico di Cagliari, Via della Scienza 5, 09047 Selargius (CA), Italy
[18]National Centre for Radio Astrophysics, Tata Institute of Fundamental Research, Pune 411007, India.
[19]CSIRO Astronomy & Space Science, Australia Telescope National Facility, PO Box 76, Epping, NSW 1710, Australia
[20]Department of Physics and Astronomy, Oberlin College, Oberlin, OH 44074, USA
[21]Physics and Astronomy Department, University of New Mexico, 1919 Lomas Boulevard NE, Albuquerque, NM 87131-0001, USA

[22]Fakultät für Physik, Universität Bielefeld, Postfach 100131, D-33501 Bielefeld, Germany

E-mail: tdolch@hillsdale.edu

Abstract. Dense, continuous pulsar timing observations over a 24-hr period provide a method for probing intermediate gravitational wave (GW) frequencies from 10 microhertz to 20 millihertz. The European Pulsar Timing Array (EPTA), the North American Nanohertz Observatory for Gravitational Waves (NANOGrav), the Parkes Pulsar Timing Array (PPTA), and the combined International Pulsar Timing Array (IPTA) all use millisecond pulsar observations to detect or constrain GWs typically at nanohertz frequencies. In the case of the IPTA's nine-telescope 24-Hour Global Campaign on millisecond pulsar J1713+0747, GW limits in the intermediate frequency regime can be produced. The negligible change in dispersion measure during the observation minimizes red noise in the timing residuals, constraining any contributions from GWs due to individual sources. At 10^{-5} Hz, the 95% upper limit on strain is 10^{-11} for GW sources in the pulsar's direction.

1. Introduction

The direct detection of gravitational waves (GWs) is a goal for which pulsar timing arrays (PTAs) are well-suited. PTAs are collections of millisecond pulsars (MSPs) distributed throughout the sky that are regularly monitored with large, typically single-dish radio telescopes. MSPs are recycled neutron stars (NSs) with low spin-down rates. The rotational regularity of MSPs means that their beamed radio emission functions as a stable clock. Deviations from regularity may signify a GW-induced distortion in the spacetime metric, changing the apparent distance between pulses and thus changing the observed spin frequency of a pulsar. PTA-oriented scientific collaborations around the world (the EPTA collaboration [1], NANOGrav [2], and the PPTA collaboration [3]; working together as the IPTA collaboration [4]) have been regularly monitoring their respective sets of pulsars for over a decade. The best-case GW detection would show the correlated signature between the pulse times-of-arrival (TOAs) from different pulsars predicted in [5]. Such a signal would be free of spurious effects along a particular line-of-sight (LoS), such as long-term, stochastic rotational fluctuations in a particular pulsar, also known as timing noise or spin noise. Although GW detection has not yet occurred with PTAs, constraints on galaxy evolution models have already been obtained [6, 7, 8], in some cases limiting the presence of GWs using multiple-pulsar data, and in others using data from a single pulsar. These non-detection limits are at frequencies near a nanohertz.

2. Data Reduction, Pipeline, and Limits on Individual Sources

The PSR J1713+0747 24-hour global campaign [9] was a day-long, continuous observation using nine radio telescopes on 22-Jun 2013 (MJD 56465 – 56466). PSR J1713+0747 was chosen because of the low root-mean-squared (rms) timing residuals with respect to its timing model, and because it is regularly observed by all the PTA collaborations. The Arecibo, Effelsberg, GMRT, Green Bank, LOFAR, Lovell, Nançay, Parkes, and WSRT radio telescopes were all used for the observation. For the much longer datasets usually used in PTAs, red noise in dispersion measure (DM) variations propagates into timing residuals [10]. DM is the quantity that measures the integrated column density of free electrons along the LoS to a pulsar. The 24-hr campaign obtained high signal-to-noise (S/N) TOAs at 1.4 GHz, showing that the timing residuals were free of any red noise due to DM variations. Noise in the residuals due to interstellar scintillation (ISS) was also modeled and shown to be white. This procedure was performed in [9] using the frequency-dependent (FD) parameters [11] that modeled the pulse profile's evolution with frequency. Otherwise, performing a weighted sum will generate non-white noise structures due the the fact that ISS creates a varying S/N across the observing band. In that case, the varying S/N would result in different TOAs in each sub-band due to the intrinsic profile's changing shape with frequency.

The timing residuals across the varying observation bandwidths are shown as a weighted sum, which we refer to here as the "broadband" residuals, in Fig. 1. Around 8 hr, ISS was occurring, which decreased the S/N. The lower S/N ratio was in spite of the fact that the GMRT's sensitivity is comparable to that of other telescopes. These otherwise high-S/N data motivate a search for high-frequency GWs from individual (or continuous wave; CW) merging supermassive black hole binaries (SMBHBs), rather than a stochastic background of GWs from such sources from 10^{-9} – 10^{-7} Hz. For the 24-hr campaign, we are sensitive to GWs at 10^{-5} – 10^{-3} Hz. At these high GW frequencies, we do not expect sensitivity to the GW background because of its powerlaw spectrum, which dominates at low frequencies. However, a nearby SMBHB with a sufficient chirp-mass could produce CWs. [12] also demonstrates that the 10^{-5} – 10^{-3} Hz range may be a good discovery band for unanticipated, periodic GW sources. We use the same detection procedure for sinusoidal GWs as [13], where details of the pipeline can be found. The only difference between the pipeline as typically used for a PTA dataset and for the present dataset is simply that only one pulsar, PSR J1713+0747, is included in the array. [13] contains several detection algorithms; here we utilize the frequentist, fixed-noise model, given that we have already established the white-noise character of the dataset. The pipeline accounts for the uneven sampling and the varying S/N of the data over time. An additional pulse phase term due to any GW perturbations on the pulsar itself is included.

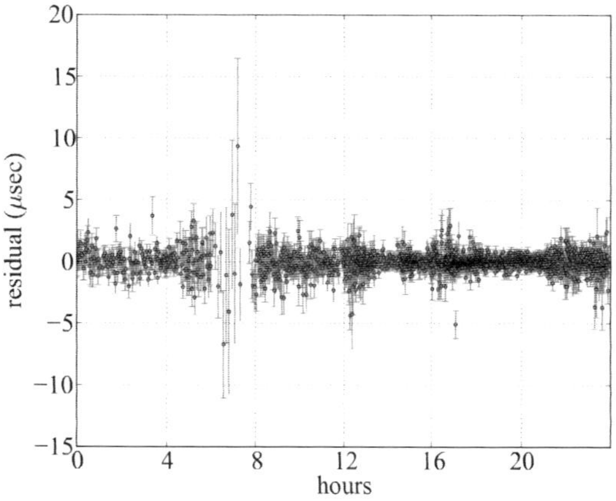

Figure 1. The 1.4 GHz broadband, 8-telescope timing residuals for the 24-hour global campaign on pulsar J1713+0747. The white-noise character of these residuals motives a search for GW in the frequency range 10^{-5} – 10^{-3} Hz. The integration time is 120 s, the variation in rms is due to ISS and changing observation bandwidth, and the error bars are the $1\,\sigma$ residual uncertainty.

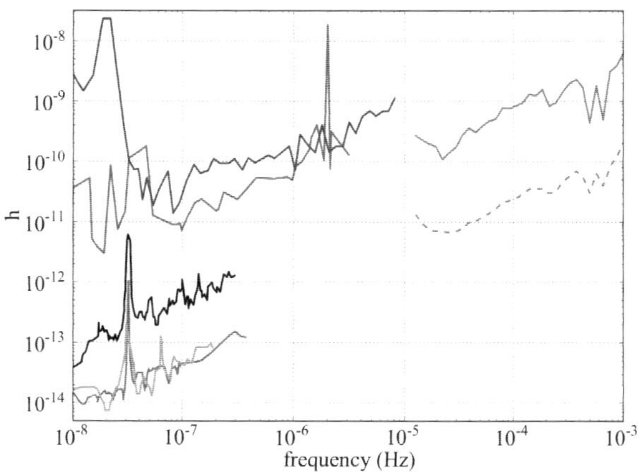

Figure 2. GW limits from the J1713 24-hr global campaign (red) from individual GW sources in random sky directions. The dashed curve shows the GW limit in the direction of the pulsar. The other CW limits shown are: pulsar J0437 [14] (blue), pulsar B1937 [15] (magenta), NANOGrav [13] (black), the EPTA [16] (green), and the PPTA [17] (cyan).

We obtain two 95% upper limit non-detection curves (Fig. 2) for the 24-hour global data: one for individual sources in a random direction in the sky (solid red curve), and the other for sources in the direction of PSR J1713+0747 itself (dashed red curve). Both show limits in GW strain h as a function of GW frequency ($f_{\rm GW}$). We chose a 0.01 radian cone for the directional limit. A spin-down rate fit is also included. The directional limit curve, which has $h \sim 10^{-11}$ at 10^{-5} Hz, provides about a factor of 10 improvement. Also plotted are the single-pulsar high-frequency limit curves from [14] and [15], from PSR J0437–4715 (magenta) and PSR B1937+21 (blue), respectively, for CW sources from a random sky direction. The random-direction CW limit curve from [13] (black) is also shown, even though it is derived from 17 NANOGrav pulsars rather than from a single pulsar. The curve is included because it was generated from the same CW detection pipeline; many-pulsar CW limits from the EPTA [16] (green) and the PPTA [17] (cyan), both random-direction, are also shown. We note that the NANOGrav curve can be extrapolated to meet the upper curve from the 24-hr campaign. This is consistent with the fact that PSR J1713+0747 influences the GW search procedure across all pulsars because of its exceptionally high residual S/N. Both [18] and the Cassini radio science investigation [19] also derived limits on GWs in similar $f_{\rm GW}$ bands, but these apply only to a stochastic background and we do not consider them here. There are no known GW sources near this band that are strong enough to have been detected. According to [20], the double pulsar J0737–3039 1.1 kpc away [21, 22, 23] should produce $h_{\rm max} \sim 10^{-21}$ at $f_{\rm GW} \sim 0.2$ mHz. Indeed, the current GW frequency of J0737–3039 is the highest known amongst NS binaries, and is closest to the GW frequency band considered in this work. Further considerations about other potential GW sources will be explored in a future publication. Meanwhile, we recommend further long-track studies, especially those optimized for different sky directions, or those coordinated between different telescopes monitoring different pulsars simultaneously.

3. Acknowledgments

The work of SC, JMC, PBD, TD, JAE, FJ, GJ, MTL, TJWL, DRL, DRM, MAM, NP, SMR, DRS and KS was partially supported through the NSF-PIRE program award number 0968296 and NSF-PFC award number 1430284. TD acknowledges seed-funding from the NANOGrav PFC award. PL acknowledges the support of IMPRS Bonn/Cologne and FQRNT B2. Some of this work was supported by the ERC Advanced Grant "LEAP", Grant Agreement Number 227947 (PI M. Kramer). Part of this research was carried out at the Jet Propulsion Laboratory, California Institute of Technology, under a contract with NASA. JAE acknowledges support by NASA through Einstein Fellowship grant PF4-150120. LOFAR data analysis was supported by the European Research Council under the EU's Seventh Framework Programme (FP/2007-2013) / ERC Grant Agreement #337062 (PI J. Hessels). We thank the anonymous referee.

4. References

[1] Kramer, M., & Champion, D. J. 2013, *Classical and Quantum Gravity* **30** 224009
[2] McLaughlin, M. A. 2013, *Classical and Quantum Gravity* **30** 224008
[3] Manchester, R. N., Hobbs, G., et al. 2013, *Publications of the Astronomical Society of Australia* **30** e017
[4] Manchester, R. N., & IPTA 2013, *Classical and Quantum Gravity* **30** 224010
[5] Hellings, R. W., & Downs, G. S. 1983, *Astrophysical Journal* **265** L39
[6] Shannon, R. M., Ravi, V., Coles, W. A., et al. 2013, *Science* **342** 334
[7] Lentati, L., Taylor, S. R., et al. 2015, *Monthly Notices of the Royal Astronomical Society* **453** 2576
[8] Arzoumanian, Z., Brazier, A., Burke-Spolaor, S., et al. 2015 (*Preprint* astro-ph/1508.03024)
[9] Dolch, T., Lam, M. T., Cordes, J., et al. 2014, *Astrophysical Journal* **794** 21
[10] Cordes, J. M. 2013, *Classical and Quantum Gravity* **30** 224002
[11] The NANOGrav Collaboration, Arzoumanian, Z., Brazier, A., et al. 2015, *Astrophysical Journal* **813** 65
[12] Cutler, C., Burke-Spolaor, S., Vallisneri, M., Lazio, J., & Majid, W. 2014, *Physical Review D* **89** 042003
[13] Arzoumanian, Z., Brazier, A., Burke-Spolaor, S., et al. 2014, *Astrophysical Journal* **794** 141
[14] Yardley, D. R. B., Hobbs, G. B., et al. 2010, *Monthly Notices of the Royal Astronomical Society* **407** 669
[15] Yi, S., Stappers, B. W., et al. 2014, *Monthly Notices of the Royal Astronomical Society* **445** 1245
[16] Babak, S., Petiteau, A., Sesana, A., et al. 2016, *Monthly Notices of the Royal Astronomical Society* **455** 1665
[17] Zhu, X.-J., Hobbs, G., Wen, L., et al. 2014, *Monthly Notices of the Royal Astronomical Society* **444** 3709
[18] Hui, L., McWilliams, S. T., & Yang, I.-S. 2013, *Physical Review D* **87** 084009
[19] Armstrong, J. W., Iess, L., Tortora, P., & Bertotti, B. 2003, *Astrophysical Journal* **599** 806
[20] Schutz, B. 1997, *Fundamental Physics in Space* **420** 229
[21] Burgay, M., D'Amico, N., Possenti, A., et al. 2003, *Nature* **426** 531
[22] Lyne, A. G., Burgay, M., Kramer, M., et al. 2004, *Science* **303** 1153
[23] Deller, A. T., Bailes, M., & Tingay, S. J. 2009, *Science* **323** 1327

Gravitational wave science from space

Jonathan R. Gair

Institute of Astronomy, Madingley Road, Cambridge, CB3 0HA, UK
School of Mathematics, University of Edinburgh, Edinburgh, EH9 7JD, UK

E-mail: j.gair@ed.ac.uk

Abstract. The rich millihertz gravitational wave band can only be accessed with a space-based detector. The technology for such a detector will be demonstrated by the LISA Pathfinder satellite that is due to launch this year and ESA has selected gravitational wave detection from space as the science theme to be addressed by the L3 large mission to be launched around 2034. In this article we will discuss the sources that such an instrument will observe, and how the numbers of events and precision of parameter determination are affected by modifications to the, as yet not finalised, mission design. We will also describe some of the exciting scientific applications of these observations, to astrophysics, fundamental physics and cosmology.

1. Introduction

The millihertz gravitational wave band is expected to be very rich in gravitational wave sources. Binaries in which the largest member is a supermassive black hole (SMBH) with mass around one million solar masses radiate at millihertz frequencies. Such black holes are expected to reside in the centres of less massive galaxies and play an important role in the assembly of large scale structure, but these black holes are very difficult to observe electromagnetically. In addition, binaries of stellar compact remnants in the Milky Way with periods of around one hour radiate at millihertz frequencies. Unfortunately, this rich frequency band is not accessible to ground-based instruments. Seismic noise limits the sensitivity of ground-based interferometers such as Advanced LIGO and Virgo [1] to frequencies above $\sim$ 10Hz, while pulsar timing arrays [2] are most sensitive at nanohertz frequencies and have very limited sensitivity above $\sim 1/\delta T$, where δT is the average time between consecutive observations, which is typically of the order of a week. To access this band therefore requires a later interferometer operating in space.

Proposals for a space-based gravitational wave detector have been discussed for two decades, in particular a concept for a Laser Interferometer Space Antenna (LISA) was extensively examined as a possible joint NASA/ESA project [3]. As a precursor to this, ESA commissioned a technology demonstration mission, LISA Pathfinder, to demonstrate the major technologies that will be required to make a space-based interferometer a success [4]. LISA Pathfinder will finally launch at the end of this year. The LISA project was ultimately not funded, due to NASA funding constraints. However, ESA has now selected *The Gravitational Universe* [5], a proposal for the detection of gravitational waves from space, as the science theme that will be addressed by its third large mission in this funding cycle (L3), to be launched in 2034. This science theme was based around a mission concept modified from the LISA design, which we refer to as evolved LISA (eLISA).

eLISA will consist of three satellites at the corners of an approximately equilateral triangle, in a heliocentric orbit at 1AU, but lagging behind the Earth in its orbit, and with laser beams passing in both directions between the satellites along the arms of the triangle. In the version of eLISA used to illustrate *The Gravitational Universe* proposal, the satellites were 1 million kilometres apart, the laser power was 0.7W, the telescope diameter was 25cm and the acceleration noise, that determines the instrument sensitivity at low frequencies, was taken equal to the original requirement for LISA. In addition, that eLISA configuration had only four laser links, i.e., the constellation consists of two different types of satellite – one "mother" satellite that has two lasers/telescopes and serves two arms, and two "daughter" satellites at the end of those arms, which have a single laser/telescope pointing toward the mother satellite, but no laser links with the other daughter. However, the final design of eLISA has not yet been fixed so it is important to understand how these various choices could influence the science outcome, so that the design can be optimised for maximal scientific return within the available budget. These studies are ongoing (see [6]), but we present a few initial results here. All parts of the mission design could in principle be varied, but we will only consider: a) the effect of changing from four links to six links, i.e., having laser links passing along two arms or three arms; b) the effect of varying the arm length from 1Gm to 2Gm or 5Gm; and c) the effect of changing the acceleration noise from the LISA design requirement to ten times worse than that. We change the telescope diameter and laser power in conjunction with changes (b) and (c), in such a way that the sensitivity at high frequencies is kept approximately constant. Longer arm lengths do impose certain restrictions on laser power to ensure sufficient photons reach the distant satellite, however we are not restricted to the choice made here and the effect of modifying the high-frequency sensitivity should be explored in the future. The relative sensitivity of the detectors formed under modifications (b) and (c) are shown in Figure 1. The effect of (a) is to allow the construction of two independent data streams, rather than one. The sensitivity of each will be as shown in Figure 1, but having two independent data streams increases the signal-to-noise ratio by approximately a factor of $\sqrt{2}$ and helps to improve parameter estimation. We will show how these different configurations influence the scientific outcome later in this article.

In the next section we will describe the different types of source expected for space-based gravitational wave detectors, including the expected event rate and precision of parameter estimation. In Section 3 we will describe the potential scientific implications of these observations and we will finish in Section 4 with a summary.

2. Sources for space-based gravitational wave detectors

2.1. Massive black hole mergers

It is well established observationally that most galaxies host a black hole in their centre [7]. These black holes are typically very massive, with masses in the range $10^4 M_\odot$–$10^{10} M_\odot$. Following mergers between their hosts, the black holes in the centres are expected to form binaries which then inspiral and merge through the emission of gravitational waves. Less massive galaxies tend to have less massive central black holes and undergo more frequent mergers, so we expect many of these gravitational wave systems to come from binaries with total mass of $10^4 M_\odot$–$10^7 M_\odot$, which radiate in the millihertz band. The properties of these black holes are poorly constrained by electromagnetic observations, so there are several viable models that predict the observed number densities of more massive black holes, but make different predictions in the eLISA range [5, 8]. This means that eLISA will provide currently unknown information about the properties of these black holes, but also means that predictions for the number of such systems that eLISA will see are rather uncertain. Light seed models, in which black holes grow from an initial population of $O(100 M_\odot)$ black holes, predict that the baseline eLISA design would see ~ 100 events in a two year mission [9]. Heavy seed models, in which black holes grow from seeds of $O(10^5 M_\odot)$, formed from direct collapse of massive dust clouds, predict between

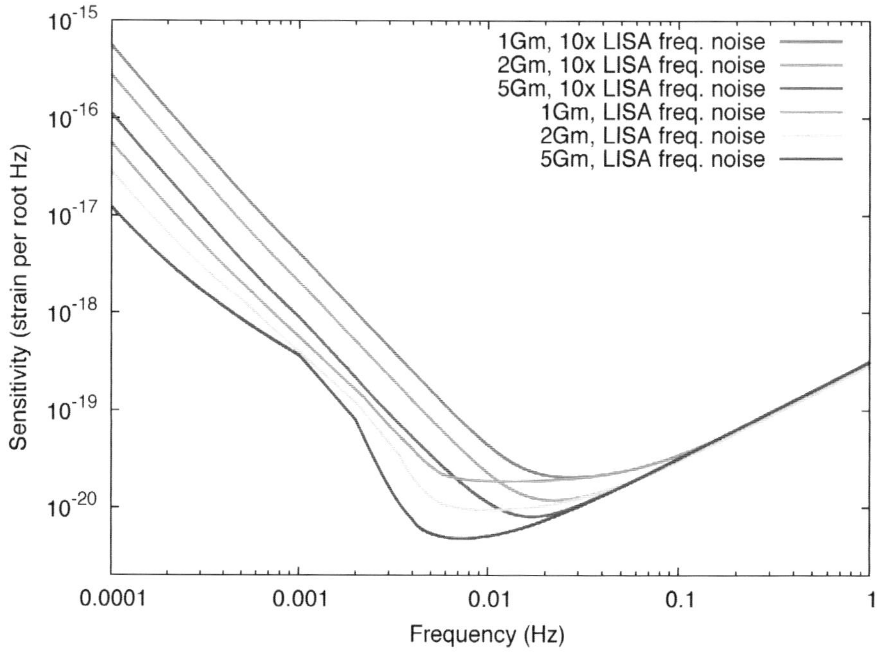

Figure 1. Sensitivity as a function of frequency for six different possible configurations of the eLISA mission. The configurations differ in arm length, which range from 1Gm to 5Gm, and in the acceleration noise performance at low frequencies, which is taken either equal to the original design specification for LISA, or ten times worse. The bump in the sensitivity curves at low frequencies is caused by the astrophysical confusion foreground from compact white dwarf binaries in the Milky Way.

a few tens of events and a few hundreds of events, depending whether it is assumed that there is a delay between the merger of the host galaxies and the formation of the black hole binary or not. Table 1 tabulates the relative gain in the expected number of events for the various alternative eLISA configurations described above. We see that increasing the arm lengths can significantly increase the number of events observed in light seed models. These extra events are primarily low-mass/high-redshift events for which the signal-to-noise ratio (S/N) is too small for them to be seen by the baseline eLISA mission [9, 10]. The arm length has little effect on event numbers for heavy seed models, in which the events are all higher mass and lower redshift and will have high S/N even in the eLISA baseline mission. Event rates for all models are significantly impacted if the acceleration noise performance is much worse than the LISA target, although these events can be partially recovered with longer arm lengths or 6-links.

Many of these SMBH binaries will be observed with high S/N, which will allow very precise determination of the parameters of the systems. A mission with the baseline eLISA design would be expected, for typical systems, to determine the masses of both black holes to between a tenth of a percent and one percent, the spins of the black holes to better than ten per cent, the sky location of the source to ~ 100 square degrees and the luminosity distance to the source to about 10% [9]. In particular, we would expect to determine both (redshifted) masses for ~ 5, ~ 15 or ~ 40 systems (for the light seed/heavy seed no delay/heavy seed with delay models respectively) and the spin of the primary for ~ 15, ~ 30 or ~ 1 systems. We would only expect, however, to measure *both* the sky position to 10 square degrees *and* the distance to 10% for one

Model	eLISA configuration					
	LISA acc. noise			10×LISA acc. noise		
	1Gm	2Gm	5Gm	1Gm	2Gm	5Gm
LS	1.00 (1.62)	2.40 (3.67)	5.25 (6.76)	0.29 (0.50)	0.73 (1.22)	1.58 (2.54)
HS with delay	1.00 (1.07)	1.06 (1.10)	1.09 (1.10)	0.32 (0.45)	0.57 (0.72)	0.89 (0.99)
HS no delay	1.00 (1.10)	1.12 (1.24)	1.26 (1.33)	0.21 (0.30)	0.41 (0.55)	0.75 (0.92)

Table 1. Change in expected number of observed SMBH merger events relative to eLISA baseline design for different mission configurations. In each cell the first number is for a 4-link configuration while the number in brackets is for a 6-link design. Each row corresponds to one of the three astrophysical models described in the text (LS = "light seed", HS = "heavy seed"). The first column (LISA acceleration noise, 1Gm arm length and 4-links) is the eLISA baseline, which would be expected to observe ~ 100, ~ 35 and ~ 450 events in the three models.

system. This increases dramatically, to about ten systems, if the configuration of the mission is changed from 4 to 6 links. Other changes to the mission design lead to more modest changes (by factors of a few up or down, following the same trend as the event rates) in this number and the number of systems with good mass and spin determinations. If systems can be well localised on the sky, there is a much better chance that the host galaxy of the source can be identified. Performing combined electromagnetic and gravitational wave observations on such sources would dramatically increase their scientific value. It is therefore clear that the single most important change to the mission design from the perspective of doing science with SMBH mergers is to upscope from 4 to 6 links, if this is possible within the budgetary constraints.

2.2. Extreme-mass-ratio inspirals

The galacto-centric massive black holes described in the previous section are typically surrounded by clusters of stars. Stars in these clusters evolve as normal and end their lives as compact stellar remnants (black holes, neutron stars or white dwarfs). Encounters with other stars can put these compact objects onto orbits that pass very close to the central black hole so that they become captured and then gradually inspiral into the SMBH via gravitational wave emission [11]. The mass ratio of these systems is typically $10^{-6} - 10^{-4}$ and so these are called extreme-mass-ratio inspirals (EMRIs). EMRIs are particularly interesting because the large mass ratio means that the small object completes a large number of orbits in the strong field region of the spacetime close to the central SMBH. The gravitational waves emitted trace out the orbit the object follows, which allows the reconstruction of a map of the spacetime structure that can be used for astrophysics and to test general relativity [12].

The intrinsic rate at which EMRIs occur in individual galaxies is very uncertain, so there is also significant uncertainty in the number of EMRI events that eLISA might observe. Inspirals of black holes are likely to completely dominate the set of observed events, since these are preferentially captured due to mass segregation and can be seen to much greater distances due to their higher intrinsic S/N. Simulations suggest that the rate should scale with black holes mass as a power law with exponent of -0.19 [13], and a Milky Way-like black hole of mass $3 \times 10^6 M_\odot$ should host ~ 400 black hole inspirals per Gyr [14]. With these assumptions and additionally assuming that the galaxies that might host EMRIs have a flat number density distribution in log mass, $\mathrm{d}n/\mathrm{d}\ln M = 0.002\mathrm{Mpc}^{-3}$, we predict the baseline eLISA mission would see ~ 650, ~ 350 or ~ 200 events depending on whether the assumed S/N threshold for detection is 15, 20 or 25 [9]. It was initially thought that EMRI detection would require a high threshold S/N due to computational limitations associated with the large parameter space that must be explored in such a search [15]. However, successful parameter estimation for EMRIs with

S/N threshold	eLISA configuration					
	LISA acc. noise			10×LISA acc. noise		
	1Gm	2Gm	5Gm	1Gm	2Gm	5Gm
15	1.00 (2.08)	3.27 (5.11)	6.55 (7.83)	0.14 (0.33)	0.60 (1.30)	1.88 (3.44)
20	1.00 (2.26)	3.93 (6.82)	9.88 (12.94)	0.12 (0.29)	0.54 (1.32)	2.04 (4.00)
25	1.00 (2.45)	4.63 (8.89)	14.29 (20.00)	0.13 (0.32)	0.55 (1.34)	2.32 (4.84)

Table 2. As Table 1 but now showing relative numbers of EMRI events expected to be observed, for three different specifications of the S/N required for detection. The eLISA baseline configuration is predicted to observe 650/350/200 events respectively for these three thresholds.

S/N less than 20 has subsequently been demonstrated in the mock LISA data challenges [16]. The latter demonstration was for a highly simplified data set, so the threshold that will be required in practice is still not known exactly. Table 2 shows how the event rate varies as the mission configuration is adjusted. The trends are similar to those seen in Table 1 for the black hole mergers. However, the EMRI rate can be much more significantly enhanced by improving the detector sensitivity, since EMRIs are sensitivity rather than astrophysical-rate limited — doubling the distance reach of the detector increases the number of observed events by a factor of 8 for EMRIs, but not for SMBH mergers, since there are no more sources in the added volume.

Despite their lower S/N, the parameters of EMRIs can be determined even more precisely than those of SMBH mergers, because any given EMRI generates so many cycles of gravitational radiation in the strong field region of the spacetime. We expect to measure the masses of both objects and the spin of the central massive black hole in any observed EMRI to a few hundredths of a percent, the sky location to a few square degrees and the luminosity distance to about ten percent [9, 16]. This is for an EMRI observed with $S/N \approx 30$ and does not depend very strongly on which mission configuration is used. More sensitive configurations will observe any given EMRI with higher S/N and hence improved parameter estimation (errors scale as $1/(S/N)$), but at fixed S/N all configurations perform equally well.

2.3. Other sources

eLISA will also observe thousands of compact binaries (primarily white dwarf-white dwarf binaries) with ~ 1 hour periods in the Milky Way. Some such systems are known to exist as they have been observed electromagnetically and these "verification binaries" can be used to check that the eLISA instrument is performing as expected. The baseline eLISA mission would be expected to observe an additional ~ 5250 individually resolved binaries with $S/N > 7$, of which ~ 1350 would be accurately resolved on the sky and of those ~ 450 would also have well determined distance [17]. Although these signals are almost monochromatic, over a two year mission with the baseline eLISA configuration, we would also expect to measure frequency derivatives for ~ 1500 systems and a second-derivative of frequency for $O(1)$ system. These numbers change in a similar way to the EMRI events (see Table 2) as the mission configuration is varied, although the number of individually resolvable events does not get above ~ 20000, since sensitivity is ultimately limited by confusion noise from unresolved compact binaries at low frequency, as indicated by the low-frequency bump in the sensitivity curves shown in Figure 1.

eLISA could also observe more exotic sources of cosmological origin, such as bursts from kinks or cusps on cosmic strings, or a stochastic background of millihertz gravitational radiation generated in the early Universe. Millihertz gravitational waves are generated at the TeV energy scale, which is the scale of the electroweak phase transition. The detection of a background at these frequencies would therefore have a profound impact on our understanding of the physics of that energy scale. We refer the reader to [18] for more details.

3. Science with space-based gravitational wave detectors

eLISA observations of the sources described above have the potential to transform our scientific understanding in several areas. We cannot go into any detail here, but in the following we will very briefly describe some of the potential scientific outcomes of eLISA observations. We refer the interested reader to [19] for a more complete description and additional references.

3.1. Astrophysics

The black holes that eLISA will observe play a vital role in the assembly of structure in the Universe. SMBH mergers trace out galactic mergers and so eLISA observations of these sources will probe the hierarchical assembly of structure. Several models exist that are consistent with current observations of higher mass systems, but these make different predictions in the low mass range that eLISA will probe. eLISA will be able to distinguish between a wide variety of different models for the growth of structure [8] and will also be able to identify when observed sources come from a mixture of different populations, and determine the mixture fraction to a precision of $\sim \pm 0.2$ [8, 9]. SMBH merger observations will also probe the accretion history of black holes, through measurements of black hole spins.

eLISA observations of EMRIs can be used to probe the number density and characteristics of massive black holes in the eLISA range in the local Universe. Observations of 10 EMRIs will be sufficient to constrain the slope of the mass function to ± 0.3, the precision with which it is currently known, and this improves as the square root of the number of events [20]. EMRI observations alone will not put constraints on the evolution of the mass function with redshift, but this might be possible by combining EMRI observations and observations of SMBH mergers. EMRIs also probe the physics of dense stellar clusters — the masses of the inspiring compact objects provide a probe of the stellar initial mass function and the efficiency of mass segregation, while the inclination and eccentricity probe the channel by which the EMRI formed.

Observations of galactic compact binaries trace stellar populations in the Milky Way, which will place indirect constraints on compact binary merger rates. Measurements of frequency derivatives in such systems also offer clues to the physics of mass transfer and of tidal interactions in stellar binaries [5, 9, 17].

3.2. Fundamental physics

eLISA observations will probe the predictions of general relativity in a regime of strong field, dynamical gravity in which it has never been tested. All future gravitational wave observations can be used for testing fundamental physics, but eLISA will provide particularly strong constraints since eLISA sources will be detected with high S/N (SMBH mergers could have S/N of thousands), they can be observed for many cycles (in particular, EMRIs can be tracked for hundreds of thousands of orbits in the strong field), the source dynamics are rich (both components will have significant spins for SMBH mergers, while EMRIs are expected to be on eccentric and inclined orbits about the central SMBH) and the systems are expected to be astrophysically "clean", i.e., the sources are mostly black holes and the influence of other material is expected to be small. There is an extensive literature describing tests of fundamental physics that will be possible with eLISA observations. We cannot describe everything here, but will give a few highlights and refer the reader to [21] for a comprehensive review.

eLISA observations can test aspects of gravitational physics, including detecting non-GR polarisations of gravitational waves [22], measuring the speed of propagation of gravitational waves through the dispersion of chirp signals [23] and detect differences in the rate at which energy is being carried away from a system which might reveal other emission channels, for instance dipole radiation in scalar-tensor theories of gravity [24]. It will also be possible to place constraints on generic deviations from the predictions of general relativity, by comparing measured values of the post-Newtonian coefficients of an expansion of the phase evolution as a function of frequency with the values predicted by GR [25]. Deviations of a tenth of a percent in low order coefficients should be identifiable. Alternatively, generic additional terms can be included in the phase expansion and constrained by observations (the parameterised post Einsteinian formalism [26]). eLISA will be able to exclude a large part of the ppE parameter space, mostly at higher orders in frequency that are not probed by binary systems where velocities are typically much smaller [27].

eLISA observations of EMRIs will probe the nature of black hole spacetimes. General relativity predicts that a black hole is completely characterised by two parameters — a mass and a spin. All higher terms in an expansion of the gravitational field in powers of distance are determined by the mass and spin. The waveforms generated by EMRIs encode a map of the spacetime structure from which those higher terms can be measured and compared to the predictions [12]. Spacetime mapping can also be done using eLISA measurements of the modes of the ringdown radiation from SMBH mergers [28]. Spacetime mapping with EMRIs has been explored extensively (see [21] for a full review). eLISA's ability to measure "bumpiness" of black holes was characterised using various different models for the bumps, e.g., [29, 30, 31, 32]. The conclusion was that typical EMRI observations would simultaneously measure the central black hole mass and spin to a hundredth of a percent and constrain any deviations in the quadrupole moment at the level of a tenth of a percent [33]. EMRI observations could also indicate the presence of a horizon in the spacetime [34], reveal the physics of the tidal coupling interaction between the two bodies [35] and detect the influence of perturbing matter [36] or stellar objects [37]. These astrophysical imprints are distinct from those of the central object.

3.3. Cosmology

The direct detection of a millihertz gravitational wave background would provide constraints on the physics of the early Universe at the TeV scale, as mentioned earlier. However, eLISA sources can also be used indirectly to probe the expansion history of the Universe. SMBH mergers and EMRIs will both provide reasonably accurate measurements of luminosity distance, but to probe the expansion history we also need an estimate of redshift. Redshift only enters the gravitational wave signal in the form of a redshifted mass, but the intrinsic mass is unknown. The redshift could be determined if an electromagnetic counterpart to a source was observed, but the large error boxes make this unlikely. A more promising approach is to use a statistical technique, averaging over many events. If the EMRI event rate is as high as predicted, EMRIs could provide a measurement of the Hubble constant. A $1\% - 2\%$ measurement would be possible using 20 EMRI events at redshift $z < 5$ [38]. Applying similar statistical techniques to SMBH mergers could provide a probe of the equation of state of dark energy [39], if sufficient systems are observed. Any cosmological constraints that eLISA provides are unlikely to be better than those which will have been derived electromagnetically by that time. However, the eLISA measurements are completely independent of all other techniques and so will provide a completely independent verification of previous results.

4. Summary

The millihertz gravitational wave band is rich in sources and will be finally opened up by the ESA L3 mission in 2034. This mission is expected to observe gravitational waves from the mergers of binaries of massive black holes with mass in the range $10^4 - 10^7 M\odot$, from the extreme-mass-ratio inspirals of stellar compact remnants into such black holes, from hour-period compact object binaries in the Milky Way and possibly a cosmological background or bursts from cusps on cosmic strings. The design of the eLISA satellite that will constitute this mission has not yet been finalised and so work is underway to explore the impact of design choices on the potential scientific impact of the mission. Increasing the arm length increases the expected number of observed events for all source types, as does increasing the number of laser links in the constellation from 4 to 6. However, if the acceleration noise at low frequencies is significantly worse than the design target it will reduce the number of sources. The precision of parameter estimation for most sources responds in a similar way to the number of events under these design modifications, with the exception of the precision of localisation (on the sky and in distance) of SMBH merger events. For those events, there is a dramatic improvement when the number of links is increased. Therefore, of all the modifications that might be achievable within the budgetary constraints, increasing the number of links is likely to have the biggest impact on the science. Whatever the final design, eLISA observations have the potential to revolutionise our understanding of astrophysics, fundamental physics and cosmology. They will provide unique constraints on the hierarchical assembly of large scale structure, they will provide precise tests of the predictions of general relativity in a previously unexplored regime, they will provide verification of the nature of the spacetime exterior to black holes and will place new independent constraints on the expansion history of the Universe. This science will only be realised with appropriate design of the mission and careful development of techniques for the analysis of the data. There is still much work to be done between now and the L3 launch to prepare the way for the advent of millihertz gravitational wave astronomy!

References

[1] Aasi J et al. 2013 *Preprint* arxiv:1304.0670
[2] Hobbs G et al. 2010 *Class. Quantum Grav.* **27** 084013
[3] Danzmann K et al. 1996 *Class. Quantum Grav.* **13** 247
[4] Armano M et al. 2009 *Class. Quant. Grav.* **26** 094001
[5] The eLISA Consortium 2013 The Gravitational Universe *Preprint* 1375.5720
[6] Klein A et al. 2016 *Phys. Rev.* D **93** 024003
[7] Croton D J et al. 2006 *Mon. Not. Roy. Astron. Soc.* **365** 11
[8] Sesana A et al. 2011 *Phys. Rev.* D **83** 044036
[9] Amaro-Seoane P et al. 2013 *GW Notes* **6**
[10] Arun K G et al. 2009 *Class. Quantum Grav.* **26** 4027
[11] Amaro-Seoane P et al. 2007 *Class. Quantum Grav.* **24** R113
[12] Ryan F D 1995 *Phys. Rev.* D **52** 5707
[13] Amaro-Seoane P and Preto M 2011 *Class. Quantum Grav.* **28** 094017
[14] Hopman C 2009 *Class. Quantum Grav.* **26** 094028
[15] Gair J R et al. 2004 *Class. Quantum Grav.* **21** S1595
[16] Babak S et al. 2009 *Class. Quantum Grav.* **27** 084009
[17] S. Nissanke et al., Astrophys. J. **758** 131 (2012).
[18] Binetruy P et al. 2012 *JCAP* **6** 27
[19] Gair J R 2015 *Astrophysics and Space Science Proc.* **20** 225
[20] Gair J R et al. 2012 *Phys. Rev.* D **81** 104014
[21] Gair J R et al. 2014 *Liv. Rev. Rel.* **16** 7
[22] Tinto M and da Silva Alves M E 2010 *Phys. Rev.* D **82** 122003
[23] Berti E et al. 2011 *Phys. Rev.* D **84** 101501
[24] Yagi K and Tanaka T 2010 *Phys. Rev.* D **81** 064008
[25] Arun K G et al. 2006 *Class. Quantum Grav.* **23** L37
[26] Yunes N and Pretorius F 2009 *Phys. Rev.* D **80** 122003
[27] Cornish N et al. 2011 *Phys. Rev.* D **84** 062003
[28] Gossan S et al. 2012 *Phys. Rev.* D **85** 124056
[29] Collins N A and Hughes S A 2004 *Phys. Rev.* D **69** 124022
[30] Glampedakis K and BabakS 2008 *Class. Quantum Grav.* **23** 4167
[31] Gair J R et al. 2008 *Phys. Rev.* D **77** 024035
[32] Canizares P et al. 2012 Phys. Rev. D **86** 044010
[33] Barack L and Cutler C 2007 *Phys. Rev.* D **75** 042003
[34] Kesden et al. 2005 *Phys. Rev.* D **71** 044015
[35] Li C and Lovelace G 2008 *Phys. Rev.* D **77** 064022
[36] Barausse E et al. 2014 *Phys. Rev.* D **89** 104059
[37] Yunes N et al. 2011 *Phys. Rev.* D **83** 044030
[38] MacLeod C L and Hogan C J 2008 *Phys. Rev.* D **77** 3512
[39] Petiteau A et al. 2011 *Astrophys. J.* **732** 82

Computation of Schenberg response function by using finite element modelling

C Frajuca[1], F S Bortoli[1] and N S Magalhaes[2]

[1] Sao Paulo Federal Institute
Rua Pedro Vicente 625, 01109-010 Sao Paulo, SP, Brazil
[2] Federal University of Sao Paulo, Department of Exact and Earth Sciences
Rua Sao Nicolau 120, 09913-030 Diadema, SP, Brazil

E-mail: `frajuca@gmail.com`

Abstract. Schenberg is a detector of gravitational waves resonant mass type, with a central frequency of operation of 3200 Hz. Transducers located on the surface of the resonating sphere, according to a distribution half-dodecahedron, are used to monitor a strain amplitude. The development of mechanical impedance matchers that act by increasing the coupling of the transducers with the sphere is a major challenge because of the high frequency and small in size. The objective of this work is to study the Schenberg response function obtained by finite element modeling (FEM). Finnaly, the result is compared with the result of the simplified model for mass spring type system modeling verifying if that is suitable for the determination of sensitivity detector, as the conclusion the both modeling give the same results.

1. Introduction

SCHENBERG is a spherical resonant-mass Gravitational Wave (GW) detector [1, 2] being built in the Department of Materials and Mechanics of the University of Sao Paulo, Brazil, can be seen on figure 1. The sphere with 65 cm in diameter and weighting 1.15 ton, is made of a copper-aluminum alloy [3] with 94% Cu and 6% Al. The detector will have six electromechanical transducers (motion sensors that monitor the sphere surface movement, transforming the mechanical oscillation in electrical signal [4, 5]), arranged on the spheres surface in a half-dodecahedron distribution. The sensors will be located as if in the center of the six connected pentagons in a dodecahedron surface, following the studies by Merkowitz and Johnson [6, 7] confirmed by Magalhaes and collaborators [8].

By analyzing the signal of such sensors, the amplitudes and the direction of the incoming gravitation wave can be determined [9, 10]. The Schenberg group has decided to use microwave parametric transducers as motion sensors like the one used in the Australian GW detector NIOBÈ [11]. In this kind of transducers, a superconducting cavity is pumped with monochromatic resonant microwaves and when the size of the cavity changes due to vibration (one of the cavity wall is connected to the sphere by the mechanical impedance matcher), two side bands are created in the microwave signal that leaves the cavity. The amplitude of the side band is proportional to the amplitude of the sphere vibration. Such transducer was tested in SCHENBERG [12]. The history of the decisions process that has determined Schenberg characteristics can be seen in [13].

External seismic noise can degraded the detector sensitivity, to decrease such noise the detector will have a suspension (Figure 2) that has the responsibility to decrease this noise bellow the thermal noise of the sphere, in such a way, Schenberg is a vibrational system as the output being measured is the vibration of the sphere surface.

The simulation was done using SolidWorks Simulation 2010-2011 version.

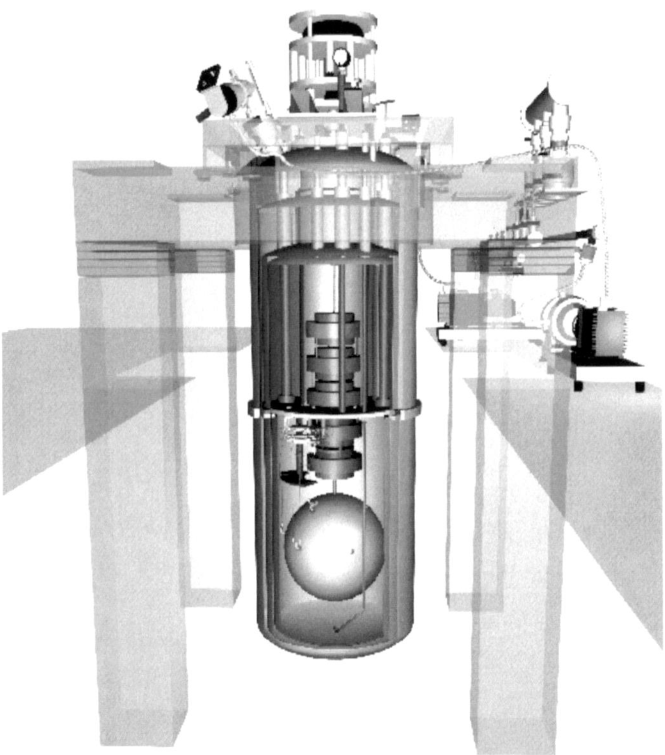

Figure 1. Schenberg schematics.

In this work, the system is simulated connecting one impedance matcher on the sphere surface then a white force is applied in the opposite side of the sphere and the frequency response is obtained. Then the frequency response graphic is compared with the results expected from a simplified theoretical model that compares the detector with a system of spring and masses, usually used in the analyses of such detectors. Using of the experience gained with the work with the simulations of the mechanical part of the transducers connected to the sphere surface using finite element modelling.

2. Material used

The materials used in the FEM simulation are summarized in the table 1.

3. The sphere by itself

First of all the sphere was simulated by itself. The values of the frequencies found were (in Hz) 3157.87; 3157.95 ; 3158.06 ; 3158.21 and 3158.32 Hz. These values have a mean of 3158.08 Hz with a standard deviation of about 0.1%. The snapshots sides of the ball movement can be in

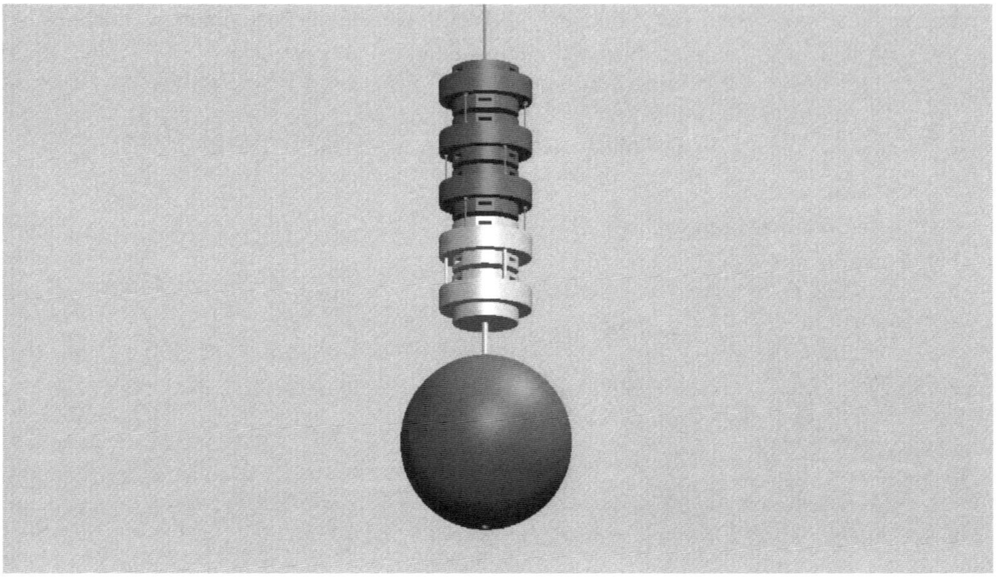

Figure 2. Schenberg suspension connected to the sphere [14].

Table 1. Material properties.

Properties	Symbol	Value 4
Material	-	94%Cu-6%Al
Specific mass	ρ	8077.5 kg/m^3
Young modulus	E	1.303x10^{11} Pa
Poisson ratio	ν	0.364
Damping ratio	-	0.0005

figure 3. The sphere oscillates from (a) to (b) then from (b) to (c) and then from (c) to (a). The loss of the five-fold quadrupole degeneracy is due to the finite element modeling approximation. If the sphere have a central hole for the suspension the frequencies are: 3172.50; 3183.00; 3213.60; 3222.90 and 3240.00 Hz. In the second case the central hole breaks the spherical symmetry.

4. The mechanical impedance matcher

The shape chosen for the mechanical impedance matcher was one that is the easiest possible to have its frequency changed into the finite element program: mushroom shape. This shape can be seen in the figure 4 together with its normam modes frequencies calculated by the FEM on the sphere surface. While this does not address the actual format to be used in the detector, its engagement with the antenna is physically similar, which can be simulated by masses and springs.

For simplicity, only one transducer was used in this simulation set of simulations. To correct tune the correct frequency of the impedance matcher to that of the sphere, their dimensions were changed until the resulting scattered modes be as close as possible. The original estimate was based on the frequency described in the book of natural shapes and frequencies by Blevins

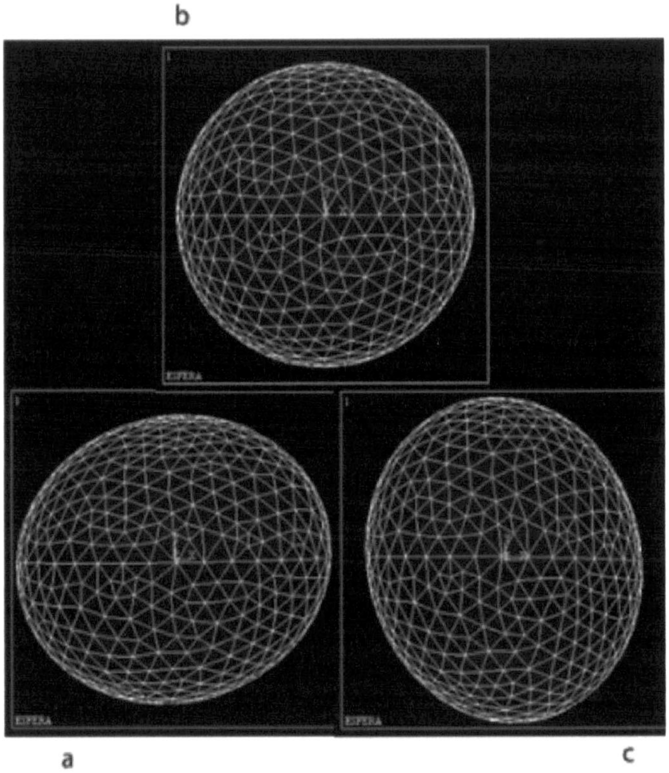

Figure 3. Simulation of the sphere by itself.

[14]. The dimensions of the impedance matcher are: 50 mm of external diameter, 10 mm of base diameter and 1.615 mm of thickness. It is made of the same material of the sphere.

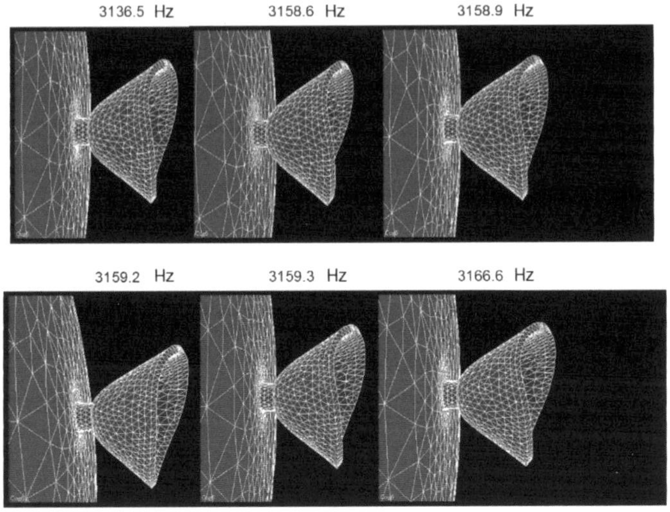

Figure 4. Modes of the impedance matcher on the sphere surface

The five modes showed in figure 4 are located in a bandwidth of 30.1 Hertz. The model to compute this bandwidth can be seen in [15]. In order to compute this it is necessary the effective mass of the sphere [16], that gives 313.4 Kg and the effective mass of the impedance matcher [17] what gives 20 g. The calculated bandwidth is of the order of 50 Hz.

5. The FEM noise spectral density

The frequency response due to a white force applied to the surface of the sphere in the opposite side is show in Figure 5. The upper curve is the deformation in the impedance matcher edge (which has the biggest amplitude) and the lower curve is the deformation in the region where the white force of 1 N^2/Hz is applied, in the region close to 3960 Hz, a peak due to tangencial quadrupole mode can be seen and in the region close to 3160 Hz the radial quadrupole mode can be seen. Figure 6 is the same curve closer to the central frequency, it is much easier to estimate the bandwidth in the amplified figure. The bandwidth is around 30 Hz and the amplification is around 10,000 times (length squared).

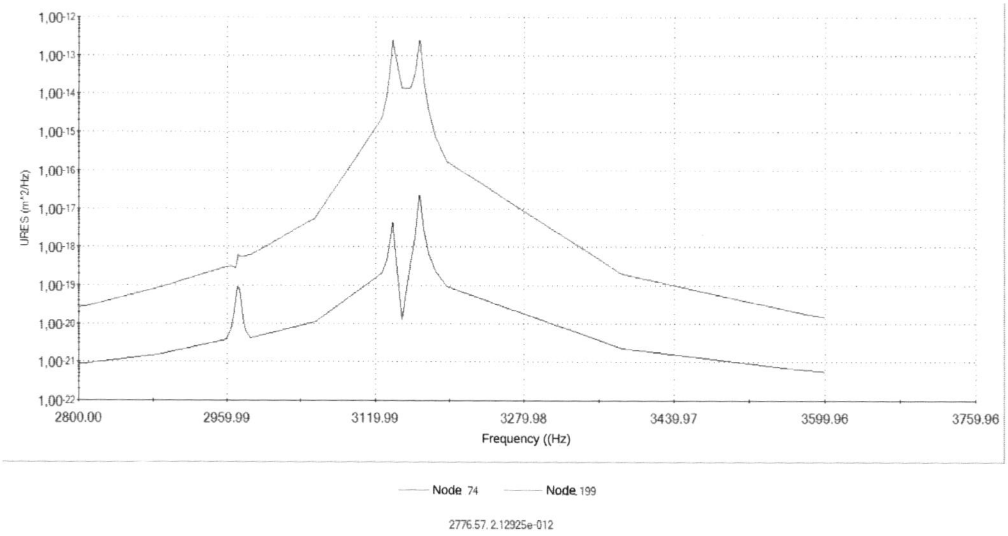

Figure 5. The frequency response on the edge of the impedance matcher due to a white force applied to the white of the sphere.

6. Conclusions

The amplification of the vibration, that can be seen as the diference of the curve in figure 5 in the peak around frequency of 3166 Hz is 100 times (considering the length, the figure shows the length squared over Hertz), in the peak around 3136 Hz the amplification, in length, is higher around 300 times. It should be, by the mathematical model equal to the squareroot of the ratio between the effective masses, which is of the order of 125.

For the the range of the two which is at the order of 30 Hz in figure 5, which is around 105 times smaller than the mean frequency. The bandwidth computed to be 50 Hz, an interesting feature is that the frequency range in figure 5 in the valley between the two peaks is almost 50 Hz. These diferences and the one of the amplification factor are compatibles, it could be improved if the steps in the frequencies were smaller, giving a better resolution for the frequency response curve, the next step of this research is to minimize these steps and improve the output.

Another interesting factor is that the amplification and the ratio between mean frequency and the bandwidth are compatible, as for theory in [15], both should be the squareroot of the effective masses which gives 125. So the FEM simulation presents compatible results of the one predicted by the modelling using springs and masses, confirming both approaches.

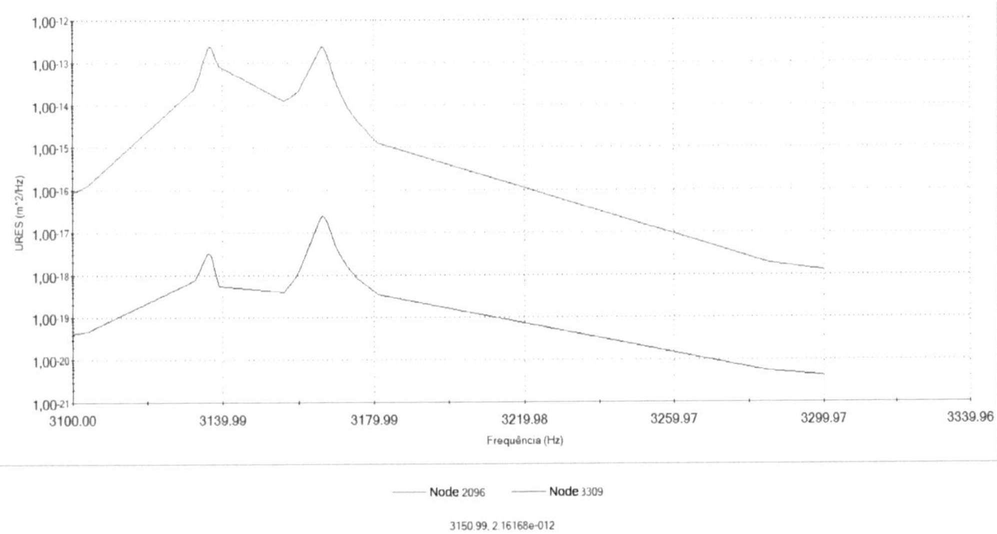

Figure 6. The frequency response closer to the central frequency.

Acknowledgments
CF acknowledges the Brazilian funding agencies FAPESP (grant # 2013/26258-4) and CNPq (grant # 312906/2013-7) for the financial support.

References
[1] Aguiar O D et al 2012 *J. Phys.: Conf. Ser.*, **363**(1) 012003
[2] Frajuca C, Ribeiro K L, Aguiar O D, Andrade L A, Castro P J, Magalhaes N S and Marinho Jr R M 2004 *CQG* **21** 1107
[3] Frossati G 1997 *Proceedings of the First International Workshop for an Omnidirectional Gravitational Radiation Observatory*, Velloso, Jr. W F, Aguiar O D and Magalhaes N S, editors (Singapore, World Scientific)
[4] Frajuca C, Ribeiro K L, Andrade L A, Velloso Jr W F, Melo J L, Aguiar O D and Magalhaes N S 2002 *CQG* **19** 1961
[5] Frajuca C, Aguiar O D, Magalhaes N S, Ribeiro K L and Andrade L A 1999 *Proc. 3rd Edoardo Amaldi Conference on Gravitational Waves (Pasadena, USA, July 1999) and AIP Conf. Proc.* **523** (New York, AIP) p. 417.
[6] Johnson W W and Merkowitz S M 1993 *Phys. Rev. Lett.* **70** 2367
[7] Merkowitz S M and Johnson W W 1997 *Phys. Rev. D* **56** 7513
[8] Magalhaes S M, Aguiar O D, Johnson W W and Frajuca C 1997 *Gen. Relat. Grav.* **29** 1511
[9] Magalhaes N S, Johnson W W, Frajuca C and Aguiar O D 1995 *MNRAS* **274** 670
[10] Magalhaes N S, Jonhson W W, Frajuca C and Aguiar O D 1997 *Astrophysical Journal* **475** 462
[11] Blair D G, Ivanov E N, Tobar M E, Turner P J, van Kann F and Heng I S 1995 *Phys. Rev. Lett.* **74** 1908
[12] Aguiar O D et al 2006 *CQG* **25** 114042
[13] Aguiar O D et al 2002 *Braz. J. Phys.* **32** 4 p. 866 (Dec)
[14] Blevins D 1984 *Formulas For Natural Frequency and Mode Shape* (Malabar, fla.: Krieger publ.) c1979
[15] Richard J P 1979 *Proceedings of Second Marcel Grossmann Meeting on General Relativity* ed. Rufini R (North-Holland, Amsterdam)
[16] S. M. Merkowitz and W. W. Johnson 1995 *Phys. Rev. D*, **51** 2546
[17] Ogawa Y and Rapagnani P 1984 *Il Nuovo Cimento C* **7** 21

Journal of Physics: Conference Series **716** (2016) 012017 doi:10.1088/1742-6596/716/1/012017

Status of the cryogenic payload system for the KAGRA detector

R Kumar[1], D Chen[2], A Hagiwara[1], T Kajita[2], T Miyamoto[2], T Suzuki[1], Y Sakakibara[2], H Tanaka[2], K Yamamoto[2], and T Tomaru[1]

[1]High Energy Accelerator Research Organization (KEK), 1-1 Oho, Tsukuba, Ibaraki, 305-0801, Japan

[2]Institute for Cosmic Ray Research, University of Tokyo, 5-1-5, Kashiwa-no-ha, Kashiwa, Chiba, 277-8582, Japan

E-mail: rkumar@post.kek.jp

Abstract. KAGRA is a large scale cryogenic gravitational wave telescope currently under construction in Japan. The detector is located 200 m underground in the Kamioka mine and will employ cryogenic technologies to achieve high sensitivity. The mirrors of the interferometer will be in the form of multiple pendulums and the final stages will employ cryogenic sapphire suspension system operating at 20 Kelvin. In this paper we report the ongoing activities of the cryogenic payload group involved in the design and fabrication of the cryogenic payload system for the KAGRA detector

1. Introduction

The existence of gravitational waves was proposed by Einstein in his General Theory of Relativity published in 1916 [1]. Gravitational waves are described as ripples in the curvature of space-time fabric and they interact very weakly with matter, which makes it very difficult for direct detection. The effort for directly detecting gravitational waves began in 1960 with Joseph Weber's bar detectors [2]. However, till this date there has been no success and with increased sensitivity of the instruments only the upper limits of gravitational waves signals have been defined. The first indirect evidence of gravitational waves was provided by Hulse [3] and Taylor [4] who were awarded Nobel Prize for Physics in 1993. Their discovery of binary pulsar 1913+16, along with Taylor and Weisberg's [5] estimation of the decay of orbital period with time were consistent with the predictions of general relativity.

Currently there is a global network of ground based laser interferometers being built for the first direct detection of gravitational waves. The first generation of detectors like LIGO (USA) [6], GEO 600 (Germany) [7], VIRGO (Italy) [8] and TAMA 300 (Japan) [9] were not successful in detecting gravitational waves, however they achieved very high sensitivity resulting in an upper limit of 0.015 events per day at signal level less than equal to 5×10^{-21} /•Hz [10]. The first generation of detectors are currently being upgraded to advanced LIGO, advanced Virgo and GEO-HF. There have been some new initiatives in the form of a Large-scale Cryogenic Gravitational-wave Telescope (LCGT) [11] project in Japan, also called as KAGRA. The arm length of the KAGRA detector is 3 km and is located 200 m underground in the Kamioka mine (250 km west of Tokyo). The seismic noise levels at the underground site has been measured using seismometers and is found to be in the order of 10^{-9}m/•Hz at 1 Hz, which is at least two orders of magnitude lower than other detectors around the world [12]. Another key feature of the KAGRA detector is the use of cryogenic technologies to achieve strain sensitivity of the order of $h_{rms} \sim 10^{-23}$ which is key to define the emerging field of gravitational wave astronomy. The detection range of the KAGRA detector is expected to be around 173 Mpc in full sky average, however signals from neutron star binary coalescence at a distance of 280 Mpc can also be observed at a certain detector orientation [13].

2. Current status

The KAGRA project is initiated by the Institute for Cosmic Ray Research (ICRR) at The University of Tokyo, in collaboration with the High Energy Accelerator Research Organisation (KEK) and the National Astronomical Observatory of Japan (NAOJ) in Japan. The excavation of the underground site of the detector at Kamioka began in 2012 and 7 km of tunnel work was completed in March 2014 [14]. The installation of vacuum chambers, cryostat and beam tubes began soon after and currently they are undergoing tests for detecting vacuum leakage. The KAGRA project will be operational in two phases, the first phase is iKAGRA which will operate at room temperature with a simple Michelson interferometer setup. iKAGRA is expected to be online by March 2016. The second phase is the baseline design - bKAGRA which is expected to be online by 2018. bKAGRA will employ cryogenic technologies and dual recycled Michelson interferometers with input and out mode cleaners along with 400 kW arm cavities.

3. Design of the cryogenic payload system

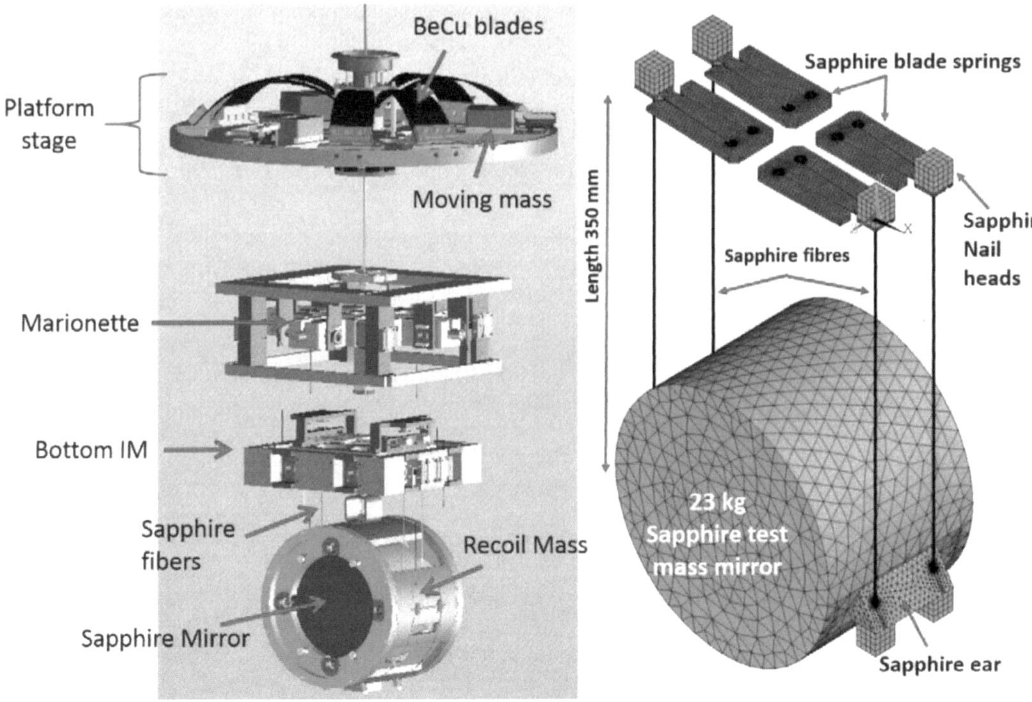

Figure 1. CAD layout of the cryogenic payload system for the KAGRA detector, figure on the right shows the FEA (ANSYS) model of the sapphire suspension system.

The mirrors of the KAGRA detector are made of sapphire. The mirror and its suspensions (called as payload) are cryogenically cooled. The payload is jointly developed by KEK and ICRR. The cryogenic payload system is suspended by room temperature multistage Vibration Isolation System (VIS) developed by NAOJ. Mirror thermal noise strongly depends upon temperature and at 20 K thermoelastic noise in the substrate (sapphire) material is nulled [15]. The cryogenic payload system will be housed in a cryostat and the mirror will be cooled to 20 K using four cryo-coolers, however the temperature at other parts of the system will be lower than 20 K for heat transfer. Figure 1 shows the CAD model of the cryogenic payload system developed for the KAGRA detector. This payload system will employ four vertical stages for seismic isolation and for achieving lower thermal noise. The uppermost is the platform stage made of stainless steel and will employ beryllium copper (BeCu) blades springs for vertical isolation. The next stage is the marionette which will control the pitch and roll movement of the mirror. The final two stages include a stainless steel intermediate mass followed by the sapphire suspension system. The 23 kg sapphire test mass mirror is suspended by four sapphire fibres of length 350 mm and diameter 1.6 mm. The fibres will be suspended using four sapphire blade springs [16]. The purpose of the blade springs is to compensate the difference in the length of the fibres having a manufacturing tolerance of 100 microns, which could result in an unequal tension between the four fibres in the suspensions. In this scenario the violin mode frequency which is currently at 175 Hz, could spread by 10%, thus reducing the bandwidth of the detector in the most sensitive region. The assembly of the sapphire parts will involve bonding of sapphire ears to the sides of the test mass. Hydroxide catalysis bonding technique [17] will be used for this purpose. The sapphire fibres have rectangular parallelepiped sapphire nail heads attached at both the ends (fabricated by IMPEX, Germany). The nail heads of the fibres will be attached to the blade springs and ears by depositing a thin layer of Indium foil using 35 W Halogen lamp [18]. The sapphire blade springs will be rigidly fixed to the intermediate mass using copper clamps.

4. Prototype test suspensions

Currently there are two types of prototype test suspensions being studied to realise and develop an assembly procedure for the final cryogenic suspension system. In KEK, parts of the prototype of the cryogenic payload system have been fabricated (shown in Figure 2(a)). In order to simplify the experiment, the sapphire components are replaced by dummy made of stainless steel. The metal prototype has beryllium copper blade springs suspending the 23 kg test mass. However, the thickness of this blade springs has been optimised to obtain spring constant and displacements values similar to the sapphire blade springs under tension.

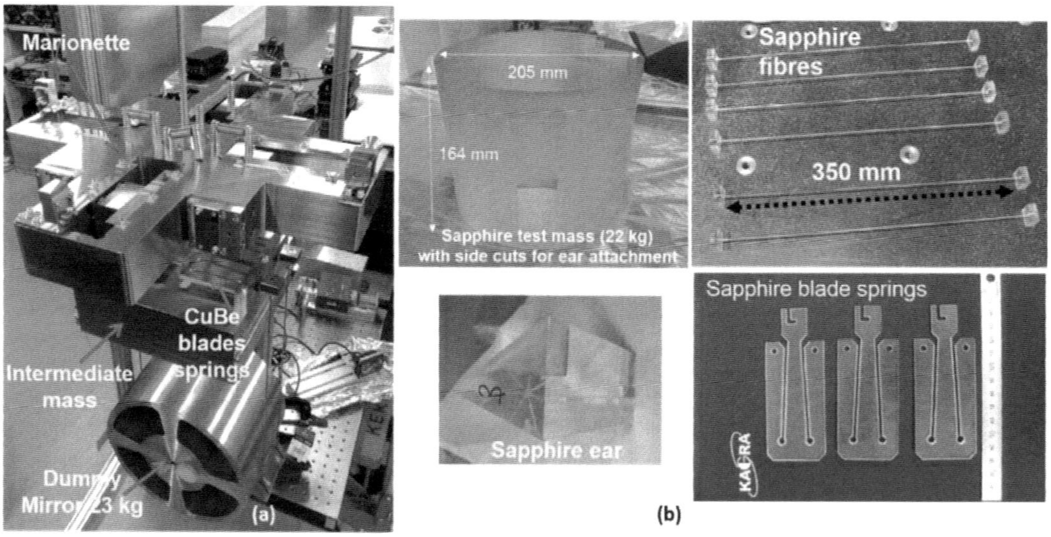

Figure 2. (a) shows the metal prototype of the cryogenic suspension system, comprising of marionette, intermediate mass with BeCu blade springs and a 23 kg dummy test mass suspended on four metal wires of length 350 mm and diameter 2.0 mm. (b) shows the sapphire components at ICRR to be used for the fabrication of prototype test sapphire suspension system.

Using this metal prototype suspension, the pitch and roll measurements of the system are currently being studied at room temperature using probe indicator and displacement meter.

A second prototype test suspensions is currently under development and will include a 22 kg sapphire mirror, four sapphire fibres, sapphire blade springs and a metal intermediate mass. For the prototype test, we have a sapphire mirror of diameter 205 mm and height 164 mm which is smaller in dimensions as compared to the real KAGRA test mass mirror (220 mm by 150 mm ~ 23 kg). The mirror has side cuts for the attachment of sapphire ears using hydroxide catalysis bonding. High quality sapphire fibres with rectangular parallelepiped nail heads have been fabricated and delivered by IMPEX, Germany. The sapphire blade springs have been fabricated by Shinkosha optics in Japan. Figure 2(b) shows the sapphire components at ICRR which will be used for the assembly of the prototype sapphire test suspensions. KEK and ICRR have the necessary facilities like cryostats and clean rooms suitable for the assembly and test of the system. The assembled system will be installed in a cryostat and cooled to 20 K and will be studied for static displacements values and thermal conductivity. The study of mechanical loss in the fibres for the fundamental violin mode (at around 175 Hz) will also be undertaken and is of utmost importance for us from thermal noise perspective. Finally, the value of mechanical loss measured experimentally will be compared with the predicted loss using FEA modelling techniques.

5. Conclusions

KAGRA is an underground detector with cryogenic technologies where the mirrors and suspensions will be cooled to 20 K. The construction, installation and testing of various components of the detector are ongoing and iKAGRA (room temperature operation) is expected to be online by March 2016. The baseline design of the KAGRA detector will be cryogenic and is expected to be online by 2018. The R&D for the cryogenic payload system for the bKAGRA is currently under progress at KEK and ICRR. The first cryogenic prototype payload system is expected to be fabricated and characterized in 2016.

Acknowledgements

This work was supported by MEXT, leading-Edge Research Infrastructure Program, JSPS Grant-in-Aid for Specially Promoted Research 26000005, JSPS Core-to-Core Program, A. Advanced Research Networks. The authors would also like to thank their colleagues in the KAGRA collaboration and at Glasgow, Rome and Jena supported by the EU under the ELiTES project (IRSES no 295153).

References

[1] Einstein A 1915 *Annalen der Physik* **49** 769
[2] Weber J 1970 *Physcial Review Letters* **25** 180
[3] Hulse R 1994 *Review of Modern Physics* **66** 699
[4] Taylor J H 1994 *Review of Modern Physcis* **66** 711
[5] Taylor J H and Weisberg J M 1982 *Astrophys. J.* **253**, 908
[6] Abott B P 2009 *Reports on Progress in Physics* **29** 7 076901
[7] Luck H et al 2006 *Classical and Quantum Gravity* **23** S71-8
[8] Accadia T et al 2012 *Journal of Instrumentation* **7** P03012
[9] Ando M and the TAMA collaboration 2005 *Classical and Quantum Gravity* **22** S881-9
[10] Abott B P et al 2007 *Classical and Quantum Gravity* **24** 5343
[11] Kuroda K et al 2010 *Classical and Quantum Gravity* **27** 084004
[12] Tomaru T 2015 *Proceedings of Monriond Conference*, Itlay
[13] Kanda N 2011 *The 11th Asian-Pacific Regional IAU Meeting*, Thailand
[14] Uchiyama T et al 2014 *Classical and Quantum Gravity* **31** 224005
[15] Kazuhiro Y et al 2006 *Physical Review D* **74** 022002
[16] Kumar R 2014 *Gravitational Wave Advanced Detector Workshop*, Takayama, Japan
[17] Douglas R et al 2014 *Classical and Quantum Gravity* **31** 045001
[18] Chen D 2015 PhD Thesis, The University of Tokyo, Japan

Measuring Gravitational-Wave Propagation Speed with Multimessenger Observations

Atsushi Nishizawa[1*] and Takashi Nakamura[2]

[1] Theoretical Astrophysics 350-17, California Institute of Technology, Pasadena, California 91125, USA

[2] Department of Physics, Kyoto University, Kyoto 606-8502, Japan

E-mail: *anishi@caltech.edu

Abstract. A measurement of gravitational wave (GW) propagation speed is one of important tests of gravity in a dynamical regime. We report a method to measure the GW propagation speed by directly comparing arrival times of GWs, neutrinos from supernovae (SN), and photons from short gamma-ray bursts (SGRB). We found that the future multimessenger observations can test the GW propagation speed with the precision of $\sim 10^{-16}$-10^{-15}, improving the previous suggestions by $9-10$ orders of magnitude. We also propose a novel method that distinguishes the true signal due to the deviation of GW speed from the speed of light and the intrinsic time delay with compact binaries at cosmological distances.

1. Introduction

The GW observations enable us to test gravity theory in strong and dynamical regimes of gravity (for reviews, [1–3]). There have been the suggestions of model-independent method to test gravity by searching for anomalous phase deviation from general relativity [4–6] and with GW polarizations [7–9]. The other test is measuring the propagation speed of a GW. In general relativity, a GW propagates with the speed of light, while in the alternative gravity theories the propagation speed could deviate from the speed of light [10–12]. Also the modification of spacetime structure due to quantum gravity effects may affect the propagation of a GW [13].

GW propagation speed has been constrained from ultra-high energy cosmic rays. Assuming the cosmic rays originate in our Galaxy, the absence of gravitational Cherenkov radiation leads to the limit, $c - v_g < 2 \times 10^{-15} c$ [14]. However, this constraint is applied to only subluminal case. So far there have been a few proposals to directly measure the GW speed. One method is comparing the phases of a GW and its electromagnetic counterpart from a periodic binary source [15, 16]. A similar method (the GW Rømer time delay) that does not rely on any EM observation was suggested recently [17]. A GW signal from a periodic GW source is modulated in phase due to the Earth revolution. In these methods, the sensitivities are limited by the baseline of the solar system and results in the measurement precision of the order of 10^{-6}.

This article is a short summary of our previous work [18], in which we have reported a simple but powerful method to measure the propagation speed of a GW by directly comparing the arrival times between GWs, and neutrinos or photons from SGRB and SN. To this end, we presuppose that SGRB is associated with a NS-NS or NS-BH binary merger [19], where NS and BH mean neutron star and black hole, respectively.

2. Method

Let us start with the comparison of the propagation speeds of a GW and a neutrino. We write the lightest neutrino mass among three mass eigenstates as m_ν and the neutrino energy as E_ν, and define the fastest propagation speed of neutrinos as v_ν. A GW is emitted at the time $t = t_e$ and is detected on the Earth at $t = t_e + T_g$, where T_g is the GW propagation time from the source to the Earth. While a neutrino is emitted at $t = t_e + \tau_{\rm int}$ with some intrinsic time delay $\tau_{\rm int}$ and is detected at $t = t_e + \tau_{\rm int} + T_\nu$, where T_ν is the neutrino propagation time. The observable is the difference of the arrival times between the GW and the neutrino. Defining $\Delta T \equiv T_\nu - T_g$, we write it $\tau_{\rm obs} = \Delta T + \tau_{\rm int}$. The first term contains the time lags due to the possible deviation of the GW propagation speed from the speed of light and the contribution of non-zero neutrino mass. The second term comes from the intrinsically delayed emission time of neutrinos at a source.

In order that the finite time lag due to the GW propagation speed different from the speed of light and neutrino mass is detectable, ΔT has to exceed uncertainties in the intrinsic time lag,

$$\Delta\tau_{\rm int} < |\Delta T| \approx T_0 |\delta_\nu - \delta_g| \,, \tag{1}$$

with $\Delta\tau_{\rm int} \equiv \tau_{\rm int,max} - \tau_{\rm int,min}$ and $\delta_\nu = m_\nu^2 c^4/(2E_\nu^2)$. At the right equality, we defined $\delta_g \equiv (c - v_g)/c$ and $\delta_\nu \equiv (c - v_\nu)/c$, and expressed ΔT in terms of δ_g and δ_ν using distance to a source L[1] and the propagation times, $T_g \equiv L/v_g$, $T_\nu \equiv L/v_\nu$, and $T_0 \equiv L/c$. For the comparison between a GW and a photon, the detectable range of δ_g is obtained by merely setting $\delta_\nu = 0$ in Eq. (1). In the derivation of Eq. (1), we have not considered the instrumental timing errors of a GW, neutrinos, and photons. However, as discussed in [18], they can be ignored because the intrinsic time delays are typically much larger.

3. Constraint on GW propagation speed

First let us focus on a SN GW-neutrino multimessenger observation. Most numerical simulations of SN with rotating progenitors [20,21] predict that GWs are mainly radiated sharply at the time of the core bounce and neutrinos are emitted within 10 msec after the core bounce. However, this is not always true for non-rotating collapses [21,22]. However, the GW waveform of the non-rotating core-collapse could easily be distinguished from that in a rotating case because of characteristic GW spikes. From this reason, we consider only the SN with rotating progenitors, in which the intrinsic time delay of neutrino emission is at most 10 msec. To find the detectable ranges of δ_g from Eqs. (1), one also need to consider an uncertainty in the neutrino mass. However, it is shown in [18] that since the neutrino mass has been constrained tightly from the cosmological observations [23], the neutrino mass can set to zero in constraining δ_g within interesting parameter ranges.

Next let us consider the intrinsic time delay of SGRB photon emission. Concentrating on the prompt emission of SGRB, high energy photons can be radiated in advance or behind the GW emission time, but this time window (of intense emission) would be less than 10 sec since the duration of the SGRB is typically less than ~ 2 sec. Thus, we use $\Delta\tau_{\rm int} = 10$ sec hereafter.

From a SN GW-neutrino event at $L = 100$ kpc, assuming $\Delta\tau_{\rm int} = 10$ msec, we have the constraint, $|\delta_g| < 9.7 \times 10^{-16}$. As for a SGRB GW-photon event, with $\Delta\tau_{\rm int} = 10$ sec and $L = 200$ Mpc, we obtain $|\delta_g| < 4.9 \times 10^{-16}$. Since the constraint on δ_g is inversely proportional to L, if SGRB is associated with NS-BH binary of mass $1.4M_\odot$ and $10M_\odot$, the distance range is ~ 3.4 times larger [24] so that the constraint would be improved by a factor ~ 3. Comparing with the previous proposals [15–17] based on direct measurements of GW propagation speed, our constraints are about 9-10 orders of magnitude tighter. We also should compare with the

[1] For cosmological sources we consider later, we must use the exact formula of the distance that takes into account the cosmic expansion.

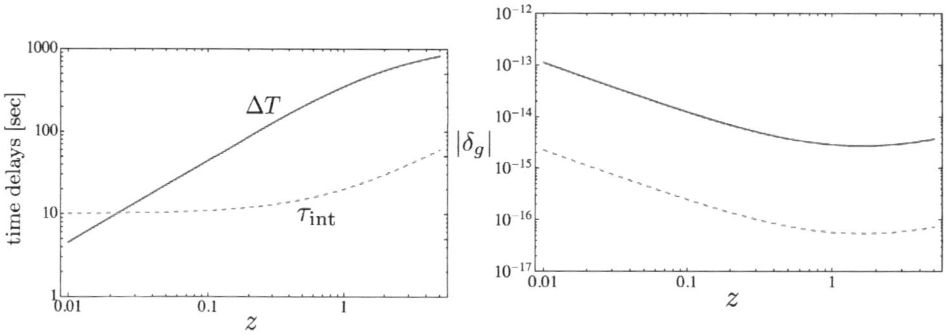

Figure 1. (left): Arrival time lag between a GW and a photon (blue, solid) and the intrinsic time delay (red, dashed) of a cosmological SGRB at redshift z when $\delta_g = 10^{-15}$ and $\tau_{\rm int}^{(e)} = 10\,{\rm sec}$. (right): Constraint on δ_g from a source at redshift z, assuming $\Delta\tau_{\rm int} = 10\,{\rm sec}$ (red, dashed) and conservatively $\Delta\tau_{\rm int} = 500\,{\rm sec}$ (blue, solid). Taken from [18].

indirect constraint on δ_g obtained so far. From the measurement of ultra-high energy cosmic rays, the constraint $0 \leq \delta_g < 2 \times 10^{-15}$ has been obtained [14]. This bound can be applied only to subluminal propagation. In this case, our method gives a stronger constraint by a factor of a few. The advantage of our method is that it is also applied to a superluminal case as well.

We comment on the event rate of coincidence detection with GW detectors such as aLIGO and neutrino detectors such as Super-KAMIOKANDE. In SN GW-neutrino observations, the coincident event rate is roughly a few events per a century [25]. In SGRB GW-γ-ray observations, the coincident event rate that has recently been estimated in [24] using BATSE data is $\sim 0.08\,{\rm yr}^{-1}$ and $\sim 3.6\,{\rm yr}^{-1}$ for NS-NS and NS-BH cases, respectively. Therefore, we can expect at least one SGRB coincident event after from a few to several years observation.

4. Using multiple cosmological SGRB

The future ground-based GW detector, Einstein Telescope (ET), extends the detection range by more than ten times and enables us to observe a million of NS-NS binaries up to $z \sim 2$ and NS-BH binaries up to $z \sim 4$ [26]. From the consideration of the beaming angle of SGRB, the number of coincidence events between GWs and SGRB photons would be more than 100 in a realistic observation time. In this section, we discuss how the time delay is affected by the cosmic expansion and how we can distinguish the time delay due to δ_g from the intrinsic time delay with multiple SGRB at cosmological distances.

The time delay induced by nonzero δ_g is given by $\Delta T = \delta_g \int_0^z H^{-1}(z')dz'$ [18], where a flat ΛCDM universe is assumed. $H(z)$ is the Hubble parameter at z. On the other hand, the intrinsic time delay at the source $\tau_{\rm int}^{(e)}$ is redshifted and is observed on the Earth as $\tau_{\rm int} = (1+z)\,\tau_{\rm int}^{(e)}$. In Fig. 1, ΔT and $\tau_{\rm int}$ are illustrated for $\delta_g = 10^{-15}$ and $\tau_{\rm int}^{(e)} = 10\,{\rm sec}$. ΔT increases at low z, proportional to the distance to the source. While, $\tau_{\rm int}$ is almost constant at low z and increases proportional to z at high z.The possible constraint on δ_g from a cosmological SGRB event becomes the tightest around $z = 1$, where the constraint is $|\delta_g| < 5.7 \times 10^{-17}$ for $\Delta\tau_{\rm int}^{(e)} = 10\,{\rm sec}$.

If multiple SGRB events observed coincidentally by GW and γ-ray detectors are available, the true signal due to δ_g can be distinguished from the intrinsic time delay at the time of emission by looking at the redshift dependence. For the data analysis, we propose a new statistics:

$$\Delta\tau_{\rm obs}(z_i, z_j) \equiv \frac{\tau_{\rm obs}(z_i)}{1+z_i} - \frac{\tau_{\rm obs}(z_j)}{1+z_j} = \frac{\Delta T(z_i)}{1+z_i} - \frac{\Delta T(z_j)}{1+z_j} + \tau_{\rm int}^{(e)}(z_i) - \tau_{\rm int}^{(e)}(z_j) \, . \tag{2}$$

Since $\tau_{\rm int}^{(e)}$ is expected to be distributed about its average, depending on a specific model of SGRB emission, the third and fourth terms would be a stochastic noise and vanishes on average. On the other hand, the first and the second terms has almost always the same signs only if signal summation is taken over the redshifts $z_j < z_i$. Therefore, this statistic effectively distinguishes the signal and the noise in our purpose detecting δ_g.

5. Conclusion

We have proposed the method to measure the propagation speed of a GW by directly comparing the arrival time lags between a GW and, neutrinos from SN or photons from SGRB. We have found that the constraint on δ_g would be 10^{-16}-10^{-15}, improving the sensitivity of the previous studies by 9-10 orders of magnitude. We also have shown that with ET one can distinguish the true signal due to the deviation of GW propagation speed and the intrinsic time delay at a source by looking at the redshift dependence.

Acknowledgments

A. N. is supported by JSPS Postdoctoral Fellowships for Research Abroad, No.25-180. T. N. is supported by Grant-in-Aid for Scientific Research, No.23540305 and No.24103006.

References

[1] Will C M 2014 (*Preprint* 1403.7377)
[2] Yunes N and Siemens X 2013 *Living Rev.Rel.* **16** 9 (*Preprint* 1304.3473)
[3] Gair J R, Vallisneri M, Larson S L and Baker J G 2013 *Living Rev.Rel.* **16** 7 (*Preprint* 1212.5575)
[4] Mishra C K, Arun K, Iyer B R and Sathyaprakash B 2010 *Phys.Rev.* **D82** 064010 (*Preprint* 1005.0304)
[5] Del Pozzo W, Veitch J and Vecchio A 2011 *Phys.Rev.* **D83** 082002 (*Preprint* 1101.1391)
[6] Yunes N and Pretorius F 2009 *Phys.Rev.* **D80** 122003 (*Preprint* 0909.3328)
[7] Nishizawa A, Taruya A, Hayama K, Kawamura S and Sakagami M a 2009 *Phys.Rev.* **D79** 082002 (*Preprint* 0903.0528)
[8] Chatziioannou K, Yunes N and Cornish N 2012 *Phys.Rev.* **D86** 022004 (*Preprint* 1204.2585)
[9] Hayama K and Nishizawa A 2013 *Phys.Rev.* **D87** 062003 (*Preprint* 1208.4596)
[10] Saltas I D, Sawicki I, Amendola L and Kunz M 2014 (*Preprint* 1406.7139)
[11] De Felice A, Nakamura T and Tanaka T 2014 *PTEP* **2014** 043E01 (*Preprint* 1304.3920)
[12] Sefiedgar A, Nozari K and Sepangi H 2011 *Phys.Lett.* **B696** 119–123 (*Preprint* 1012.1406)
[13] Amelino-Camelia G 2013 *Living Rev.Rel.* **16** 5 (*Preprint* 0806.0339)
[14] Moore G D and Nelson A E 2001 *JHEP* **0109** 023 (*Preprint* hep-ph/0106220)
[15] Larson S L and Hiscock W A 2000 *Phys.Rev.* **D61** 104008 (*Preprint* gr-qc/9912102)
[16] Cutler C, Hiscock W A and Larson S L 2003 *Phys.Rev.* **D67** 024015 (*Preprint* gr-qc/0209101)
[17] Finn L S and Romano J D 2013 *Phys.Rev.* **D88** 022001 (*Preprint* 1304.0369)
[18] Nishizawa A and Nakamura T 2014 *Phys. Rev. D* **90** 044048 (*Preprint* 1406.5544)
[19] Berger E 2013 (*Preprint* 1311.2603)
[20] Ott C, Abdikamalov E, O'Connor E, Reisswig C, Haas R *et al.* 2012 *Phys.Rev.* **D86** 024026 (*Preprint* 1204.0512)
[21] Kuroda T, Takiwaki T and Kotake K 2014 *Phys.Rev.* **D89** 044011 (*Preprint* 1304.4372)
[22] Marek A, Janka H T and Mueller E 2008 (*Preprint* 0808.4136)
[23] Zhao G B, Saito S, Percival W J, Ross A J, Montesano F *et al.* 2013 *Mon.Not.Roy.Astron.Soc.* **436** 2038–2053 (*Preprint* 1211.3741)
[24] Yonetoku D, Nakamura T, Takahashi K and Toyanago A 2014 *Astrophys.J.* **789** 65 (*Preprint* 1402.5463)
[25] Ando S, Beacom J F and Yuksel H 2005 *Phys.Rev.Lett.* **95** 171101 (*Preprint* astro-ph/0503321)
[26] Sathyaprakash B, Schutz B and Van Den Broeck C 2010 *Class.Quant.Grav.* **27** 215006 (*Preprint* 0906.4151)

How to overcome limitations of analytic solutions when determining the direction of a gravitational wave using experimental data: an example with the Schenberg detector

C F S Costa

University of Florida, Department of Physics
2001 Museum Road, 32611-8440 Gainesville, FL, USA

N S Magalhaes

Federal University of Sao Paulo, Department of Exact and Earth Sciences
Rua Sao Nicolau 120, 09913-030 Diadema, SP, Brazil

E-mail: nadja.magalhaes@unifesp.br

Abstract. It has been commonly assumed that analytic solutions can efficiently provide the direction of a gravitational wave (GW) once sufficient data is available from gravitational wave detectors. Nevertheless, we identified that such analytic solutions (based on the GW matrix reconstruction) present unforeseen theoretical and practical limitations (indeterminacies) and that for certain incoming directions they are unable to recover the latter. We present here important indeterminacy cases as well as a mathematical procedure that reduces such indeterminacies. Also, we developed a method that requires the least computational power to retrieve GW directions and which can be applied to any system of detectors able to reconstruct the GW matrix. As a test for the method, we used simulated data of the spherical, resonant-mass GW detector Schenberg, which involves five oscillating modes and six transducer readouts. The results show that this method canceled indeterminacies out satisfactorily.

1. Introduction

Determining the direction of a gravitational wave (GW) source is necessary, for instance, for confrontations between a candidate GW signal and its electromagnetic counterpart. In Figure 1 the wave's direction is z' while the detector's frame is xyz.

Gravitational waves are expected to transfer different amounts of energy to the five degenerate quadrupolar modes of the spherical antenna as a function of the incoming GW direction. In order to monitor the modes' oscillations, transducers are placed on the sphere's surface[1]. The electric readout of the k-th transducer is here described in the frequency domain by the function $I_k(f)$. We will assume that the sphere is monitored by $k = 6$ transducers. The relation between their readouts and the m-th mode of the sphere, which is described by the mode channel $h_m(f)$, is given by

$$h_m(f) = T_{mk}(f)I_k(f), \quad k = 1, .., 6, \ m = 0, 2, ..., 4 , \tag{1}$$

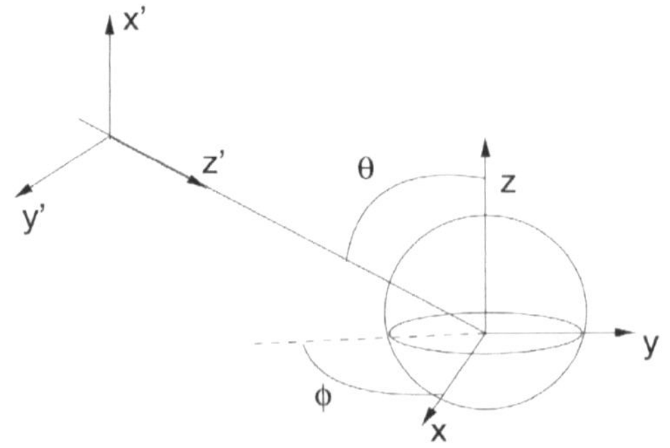

Figure 1. Position of the GW frame (primed) relative to the proper detector frame (unprimed).

where $T_{mk}(f)$ is proportional to the inverse transfer function $\mathcal{T}^{-1}(f)$ of the detector.

We can write the GW matrix in the lab frame in terms of the mode channels:

$$\mathbf{h} = \begin{pmatrix} h_0/\sqrt{3} + h_3 & h_4 & h_1 \\ h_4 & h_0/\sqrt{3} - h_3 & h_2 \\ h_1 & h_2 & -2h_0/\sqrt{3} \end{pmatrix}. \tag{2}$$

In the context of low latency pipelines it is assumed that analytic expressions (based on the GW matrix reconstruction) can efficiently provide the GW direction. In this work we indicate that this is not true in general and we propose means to overcome this limitation. Details of the results presented here can be found in [2].

2. Analytical limitations

We found that analytic expressions present theoretical and practical limitations (indeterminacies) which were not previously mentioned in the literature. Under realistic conditions, we identified that for certain incoming directions such solutions are unable to recover them. Examples of these directions are given by the red zones of Figure 2

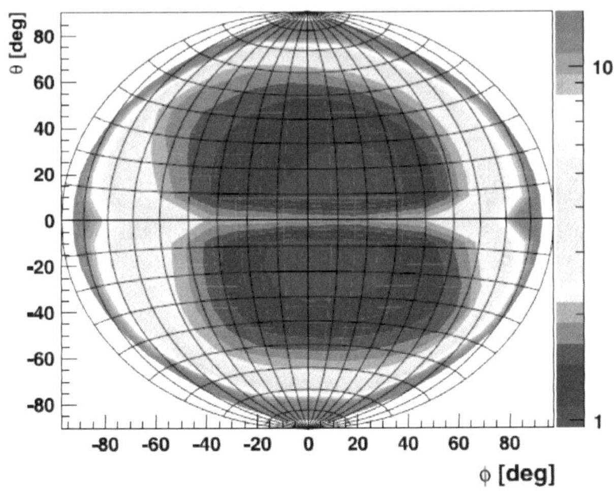

Figure 2. Aitoff projection of the average direction reconstruction errors (in degree) as functions of the injected signal directions (under SNR∼ 80). The plot shows the results when using the direction vector V_{2-3}. In the colored legend on the right of the plots (log scale in degree), values lower than two (shades of blue) indicate areas with small direction reconstruction errors in the GW direction reconstruction, while higher values (in red) indicate high direction reconstruction errors and thus indeterminacy zones. θ and ϕ are ranging from 90 to +90.

The knowledge of the five mode channels enables one to obtain the relevant quantities of the GW, particularly its incoming direction. We found that the analytic solutions depend on the

GW direction in a particular way. In Table 1 we present a summary of the different equations and their dependence on the GW direction, (θ, ϕ).

Table 1. Summary of the analytic solutions and their dependence on the incoming gravitational wave direction (θ, ϕ). The subindices $_{(1)}$ and $_{(2)}$ denote the two options to compute θ_r

Pair of lines	$1-2$	$1-3$	$2-3$
$\tan\phi_r$	$\frac{\sin\theta\cos\theta\sin\phi}{\sin\theta\cos\theta\tan\phi}$	$\frac{\sin^2\theta\sin^2\phi}{\sin^2\theta\cos\phi\sin\phi}$	$\frac{\sin^2\theta\cos\phi\sin\phi}{\sin^2\theta\cos^2\phi}$
$\tan\theta_{r\ (1)}$	$\frac{1}{\sin\phi_r}\frac{\cos\theta\sin\theta\sin\phi}{\cos\theta^2}$	$\frac{1}{\sin\phi_r}\frac{\sin^2\theta\sin^2\phi}{\sin\theta\cos\theta\sin\phi}$	$\frac{1}{\sin\phi_r}\frac{\cos\phi\sin\theta^2\sin\phi}{\cos\theta\cos\phi\sin\theta}$
$\tan\theta_{r\ (2)}$	$\frac{-1}{\cos\phi_r}\frac{\cos\theta\cos\phi\sin\theta}{\cos\theta^2}$	$\frac{-1}{\cos\phi_r}\frac{\cos\phi\sin\theta^2\sin\phi}{\cos\theta\sin\theta\sin\phi}$	$\frac{-1}{\cos\phi_r}\frac{\cos\phi^2\sin\theta^2}{\cos\theta\cos\phi\sin\theta}$

3. The proposed method

In order to avoid the limitations illustrated in Figure 2 derived from equations in Table 1 we decided to use the direction vector $\mathbf{V}$, which satisfies $(\mathbf{h} - \lambda\mathbf{I})\mathbf{V} = 0$. By using this vector we were able to reduce the indeterminacy zones to only three circles in the Aitoff projection. The vector $\mathbf{V}$ fully describes the incoming GW direction but it is more usual to give this information in terms of (θ, ϕ):

$$\cos\theta_r = V_3/|\mathbf{V}|\,, \tag{3}$$

with three options for the determination of ϕ,

$$\cos\phi_r = V_1/\sqrt{V_1^2 + V_2^2}\,, \tag{4}$$

$$\sin\phi_r = V_2/\sqrt{V_1^2 + V_2^2}\,, \tag{5}$$

$$\tan\phi_r = V_2/V_1\,. \tag{6}$$

The advantage of these solutions is that θ_r and ϕ_r are computed independently. Therefore there is no error propagation to θ_r.

We implemented our method in the context of a low latency pipeline [3] and in the presence of noise[4] applying a weighted average of the direction vector for each sample. We tested our method injecting a simulated burst into the mathematical model of the GW detector Schenberg [5, 6].

We characterize the direction reconstruction error using the angular distance

$$\delta s = \arccos\left(\sin(\theta)\sin(\theta_r)\cos(\phi - \phi_r) + \cos(\theta)\cos(\theta_r)\right)\,, \tag{7}$$

which is the angle between the direction (θ, ϕ) of the injected signal and the reconstructed direction (θ_r, ϕ_r) of the signal in the presence of noise.

We found that it is possible to cancel out indeterminacies in the presence of noise using the following method: for each direction vector $\mathbf{V_{1-2}}_t$, $\mathbf{V_{1-3}}_t$ and $\mathbf{V_{2-3}}_t$, we apply the averaging technique where each sample direction is weighted by its signal-to-noise ratio (SNR). From the three resulting averaged direction vectors $\langle\mathbf{V_{1-2}}\rangle$, $\langle\mathbf{V_{1-3}}\rangle$ and $\langle\mathbf{V_{2-3}}\rangle$, we select the direction vector that presents the lowest value of its components' standard deviation product:

$$\sigma^3(V) \equiv \sigma(V_x)\sigma(V_y)\sigma(V_z)\,. \tag{8}$$

Then using its averaged components $\langle V_x \rangle$, $\langle V_y \rangle$ and $\langle V_z \rangle$, we compute the direction in terms of the spherical coordinates (θ_r, ϕ_r).

We tested our method with the injected burst signal in the Schenberg detector and the red zones in Figure 2 turned blue, as shown in Figure 3, eliminating indeterminate incoming GW directions.

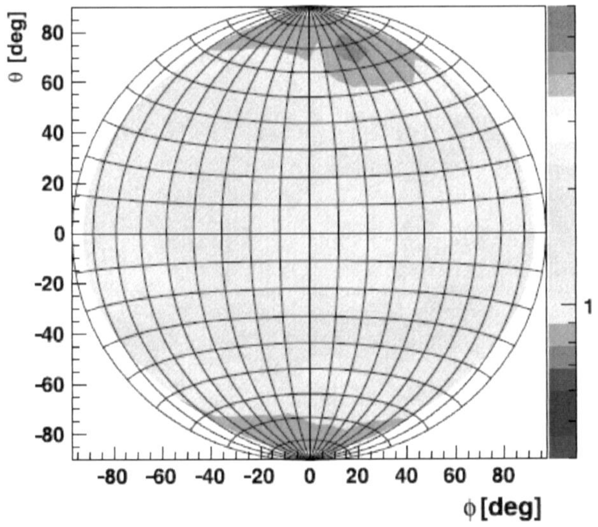

Figure 3. Aitoff projection of the average angular distances $\langle \delta s \rangle$ (in deg) as a function of the direction of the injected signal (SNR$\sim$ 80). We used our proposed method to reconstruct the GW direction in the presence of noise. The minimum is at 0.8 deg (shades of blue) and the maximum is at 1.7 deg (green).

4. Conclusions

Our method proved efficient in facing indeterminacies in the evaluation of the direction of a GW and it is suited for a low latency pipeline. It was tested with the mathematical model of the resonant-mass GW detector Schenberg, and it can be applied to any array of detectors that use the GW matrix to retrieve a wave's incoming direction.

Acknowledgments

NSM acknowledges Dr. Manuel M. M. B. de Oliveira and the Brazilian funding agency FAPESP for financial support to participate in the *11th Edoardo Amaldi Conference on Gravitational Waves* through grant # 13/26258-4.

References

[1] Frajuca C et al 2002 *CQG* **19** 1961
[2] Costa C F S and Magalhaes N S 2015 *CQG* **32** 095006
[3] Costa C A et al 2008 *CQG* **25** 184002
[4] Frajuca C et al 2004 *CQG* **21** S1107
[5] Aguiar O D et al 2012 *J. Phys.: Conf. Ser.* **363** 012003
[6] Costa C F S, Costa C A and Aguiar O D 2014 *CQG* **31** 085012

Measurement of Schumann Resonance at Kamioka

S. Atsuta[1], T. Ogawa[2], S. Yamaguchi[3], K. Hayama[4], A. Araya[2], N. Kanda[4], O. Miyakawa[5], S. Miyoki[5], A. Nishizawa[6], K. Ono[5], Y. Saito[5], K. Somiya[1], T. Uchiyama[5], M. Uyeshima[2], and K. Yano[1]

[1] Department of Physics, Tokyo Institute of Technology, 2-12-1, Oh-okayama, Meguro-ku, Tokyo, 152-8551, Japan

[2] Earthquake Research Institute, The University of Tokyo, 1-1-1, Yayoi, Bunkyo-ku, Tokyo, 113-0032, Japan

[3] Department of Biology and Geosciences, Osaka City University, 3-3-138, Sugimoto, Sumiyoshi-ku, Osaka, 558-8585, Japan

[4] Department of Physics, Osaka City University, 3-3-138, Sugimoto, Sumiyoshi-ku, Osaka, 558-8585, Japan

[5] Institute for Cosmic Ray Research, The University of Tokyo, 5-1-5, Kashiwa-no-Ha, Kashiwa City, Chiba, 277-8582, Japan

[6] Division of Physics, Mathematics, and Astronomy, California Institute of Technology, 1200 E, California Blvd, Pasadena, CA 91125, USA

E-mail: atsuta@gw.phys.titech.ac.jp

Abstract. On the stochastic gravitational-wave search, correlated noise in two or more gravitational-wave detectors can be a serious problem. Schumann resonance is the name of a standing wave of electromagnetic fields, which is one of the correlated noise sources for the second-generation gravitational-wave detectors. We measured the noise levels of the environmental magnetic field both inside and outside the mine of KAGRA site at Kamioka. In this letter, we report the result of the measurement and compare the amplitude of magnetic fields inside and outside the mine to find possible issues or gain of constructing a detector underground.

1. Introduction

One of the target sources of the second-generation gravitational-wave detectors (Advanced LIGO [1], Advanced Virgo [2], and KAGRA [3]) is the so-called stochastic gravitational-wave background (SGWB). The SGWB can be created from many uncorrelated and unresolved gravitational-waves, which contain the signals from astrophysical and cosmological events. Astrophysical sources include binary neutron stars [4], binary black holes [5], highly magnetized stars [6] and rotating neutron stars [7]. Cosmological sources are, for example, inflationary physics [8] and cosmic strings [9].

For the data analysis of gravitational-wave signals, a matched-filtering method is effective for a signal with theoretically predicted waveform [10]. The waveform of SGWB, however, cannot be predicted. A standard method for the analysis of SGWB is to cross-correlate strain data sets from two or more different detectors. Cross-correlating the data sets enables us to distinguish a signal-like noise from real gravitational waves since a noise is mostly local and is uncorrelated between different detectors.

While the first-generation detectors did not see any globally correlated noise, the sensitivity of the second-generation detectors is so high that such a small global phenomenon can create problematic correlated-noise [11]. One of the problematic global phenomena is Schumann resonance [12] [13].

Journal of Physics: Conference Series **716** (2016) 012020 doi:10.1088/1742-6596/716/1/012020

Schumann resonance is a standing wave of electromagnetic field, which occurs when a space between the surface of the Earth and the ionosphere makes a resonant cavity for electromagnetic waves. Schumann resonance is excited by lightnings that can be seen almost all over the world, and thus Schumann resonance can be seen any time from any direction. All the ground-based gravitational-detectors will then be exposed to a same Schumann resonance, except KAGRA, located underground, which might be differently affected by this phenomenon. It is important to measure Schumann resonance and compare the level inside and outside the mine.

2. Formalism

First we measure a magnetic field signal using coil sensors and collect time series data $m(t)$. We then Fourier transform the signal to obtain $\widetilde{m}(f)$ and calculate the power spectrum:

$$P(f) = \frac{2T}{N}\widetilde{m}^*(f)\widetilde{m}(f)\,. \tag{1}$$

Here N is the number of data, T is the sampling time ($T = 1/F_s$, where F_s is the sampling frequency) and t is time. In order to analyze the correlation of the signals obtained inside and outside the Kamioka mine, we also define coherence $coh(f)$:

$$coh(f) = \frac{|\widetilde{m_1}^*(f)\widetilde{m_2}(f)|^2}{|\widetilde{m_1}(f)|^2|\widetilde{m_2}(f)|^2}\,, \tag{2}$$

where $\widetilde{m_1}^*(f)$ means complex conjugate of $\widetilde{m_1}(f)$.

3. Method and Results

We used coil sensors PHOENIX GEOPHYSICS AMTC-30 magnetometers to measure the enviromental magnetic field at the KAGRA site. We set the coil sensors both inside and outside the mine with data loggers MTU-5A produced by PHOENIX GEOPHYSICS at each site. Continuous measurements with $F_s = 150$ [Hz], synchronized with GPS clocks, were performed. In a manner of the Magnetotelluric method, we placed two coil sensors in each location, one of which in the north-south (NS) and the other in the east-west (EW) directions. At the observation point outside the mine, we buried the coils about 20 cm deep in underground to avoid acoustic noise and so on. Inside the mine, as we were not able to cast a ditch, we placed and fixed the coils on the floor. The data was collected for about 30 hours. Since it was stormy and frequent spikes were found in the data set on the second observation day possibly due to frequent lightnings, we decided to use the data obtained on the first day only.

Figure 1 shows the median of the power spectrum of the magnetic field. Firstly we obtained data for 14 hours each day, and then we divided it in one-hour segments to calculate the power spectrum using DFT method. Since KAGRA was under construction during the daytime, we decided to use data obtained in the night-time to avoid noise caused by the construction work. We divided one-hour night-time data to 30-second segments and plotted the median of the power spectra.

Although Fig. 1 shows that the spectrum of the inside is higher than the outside, we do not know any clear mechanisms that promote Schumann resonance inside the mine. Several reasons can be considered for this increase: (a) calibration was not correct in some of the coil sensors, (b) the coil sensors did not work properly with the large amplitude noise, (c) the large amplitude noise raised the floor level of the spectrum. We will discuss (a) in Sec.4. We did not test (b). If (c) is the reason, a use of a certain window function in the calculation of the power spectrum would improve the problem, but we did not see any significant change.

Figure 2 shows the coherence between the signals inside and outside the mine defined in Eq.(2). Schumann resonance comes from a corresponding direction, which is determined by the direction of the

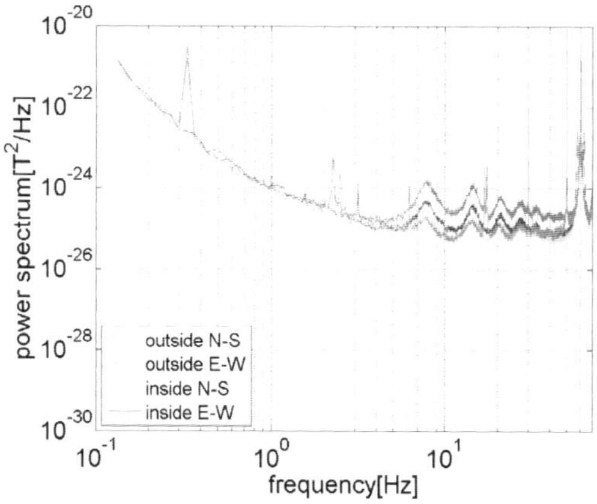

Figure 1: Median of the power spectrum of the magnetic field obtained in night-time. For all the data, Schumann resonance peaks appear at almost same frequencies (7.8 Hz, 14 Hz, $\cdots$) The peaks at 30 and 60 Hz are caused by the power line. The peak at 50 Hz appears only inside the mine, which can be from a power line for machines used inside the mine. Although some peaks were observed inside the mine, we do not refer their details since there existed a lot of unknown noise sources in the underground.

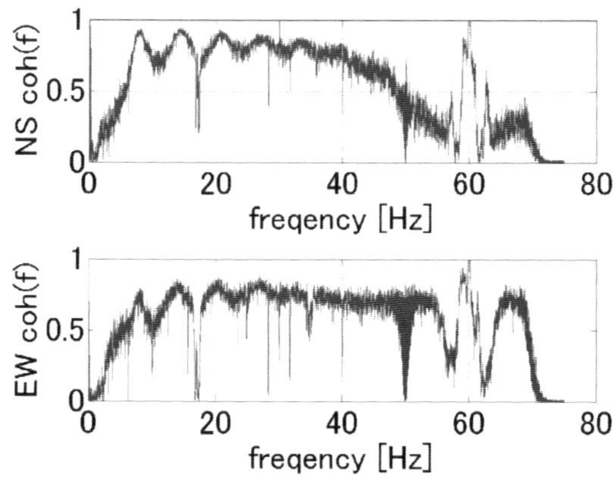

Figure 2: Coherence of the signal. At the frequency of Schumann resonance, high $coh(f)$ value indicates that the same resonance is observed both inside and outside the mine.

location of the lightning. High coherence indicates that the same Schumann resonance is observed both inside and outside the mine.

Figure 3 shows a histgram of the magnetic-field signal $m(t)$. The horizontal coordinate is the amplitude of the signal, and the vertical coordinate is the number of data with the corresponding signal. The histgram for the outside can be regarded as the Gaussian distribution with the almost zero mean, while that for the inside shows a magnetic-field signal at a certain frequency. This unknown signal has a large amplitude, which can come from the power line at 50 Hz that is seen in Fig. 1. This 50 Hz power line is much larger inside than outside. This signal could be the large signal we discussed for reasons (b) and (c).

4. Coils Test

One of the reasons we can consider for the higher level of the spectrum inside the mine may be a wrong calibration of the coil sensors. In order to check the calibration levels, we performed another measurement with these coil sensors in a magnetically quiet location: Mt. Kasagata in Hyogo prefecture.

At Mt. Kasagata, we measured a magnetic field in the same way as we did at Kamioka except for the direction of the coils. In order to measure a same signal, we placed the coils in the same direction but far from each other not to detect the magnetic signal from other coils.

Figure 4 shows the results of the simultaneous measurement. Almost no difference among the 4 coils was observed.

5. Conclusion

Schumann resonance was observed both inside and outside the Kamioka mine at the KAGRA site, and its amplitude was almost at the same level. This result means that Schumann resonance can be a problem for

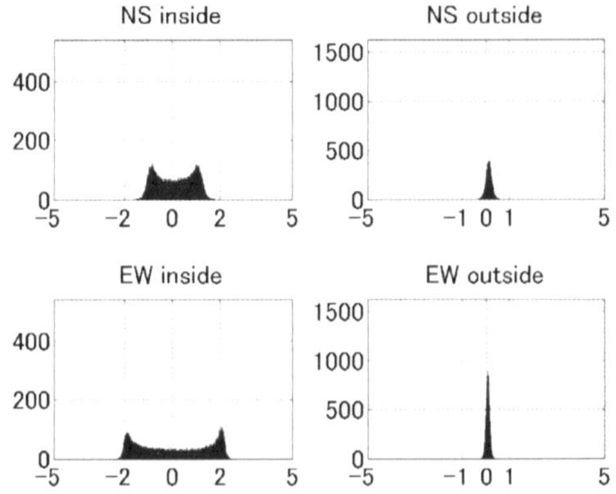

Figure 3: Histgram of each coil signal $m(t)$.The horizontal coordinate is the amplitude of the signal ($\times 10^4$), and the vertical coordinate is the number of data.

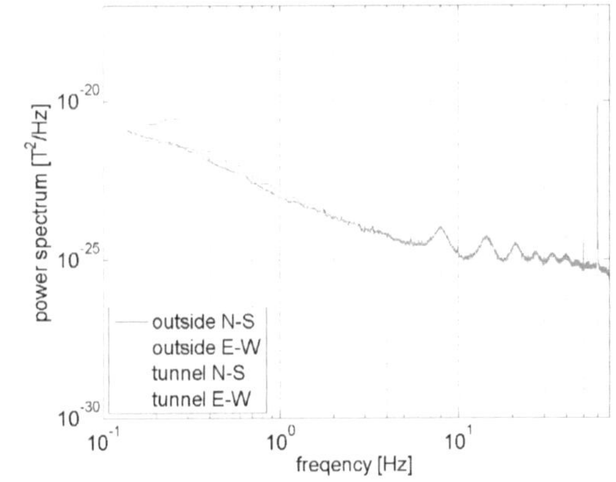

Figure 4: Magnetic median power spectrum calculated in the same way as in Fig. 1 under almost the same environment. No significant difference between coils can be seen in this figure. We concluded that the difference of calibration levels is not a problem in our measurement.

the cross-correlation method with data sets that contain KAGRA data. In fact, the power spectrum of the magnetic field inside the mine is higher than outside. We tried to find a reason for this. We checked the calibration levels by measuring the magnetic field in a quiet location but we did not see any difference. Since there were many artifacts made of iron, which has large magnetic permeability, inside the mine, they might have affected the magnetic signal obtained inside the mine.

Acknowledgments
We thank to Institute for Geothermal Sciences, Kyoto University by lending the coils we used in this letter. And we also thank to colleagues of Tokyo Institute of Technology and Osaka City University for helping our work. This work was supported by MEXT, JSPS Leading-edge Research Infrastructure Program, JSPS Grant-in-Aid for Specially Promoted Research 26000005, MEXT Grant-in-Aid for Scientific Research on Innovative Areas 24103005 and JSPS Core-to-Core Program, A. Advanced Research Networks.

Reference

[1] G. M. Harry for the LIGO Scientific Collaboration 2010 *Class. Quantum Grav.*, **27**, 084006
[2] F. Acernese et al., 2015 *Class. Quantum Grav.*, **32** 024001
[3] K. Somiya for the KAGRA collaboration, 2012 *Class. Quantum Grav.*, **29** 124007
[4] B. Allen, 2011 *Prog. Theor. Phys. Supp.*, **190**, 316
[5] X. Zhu, E. Howell, T. Refimbau, D. Blair, and Z. Zhu, 2011 *Astrophys. J.*, **739**, 86
[6] S. Marassi, R. Ciolfli, R. Schneider, L. Stella, and V. Ferrari, 2011 *Mon. Not. R. Astron. Soc.*, **411**, 2549
[7] T. Regimbau and J. A. de Freitas Pacheco, 2001 *A & A*, **376**
[8] B. Allen, 1988 *Phys. Rev. D*, **37**, 2078
[9] L. Sousa, and P. P. Avelino, 2013 *Phys. Rev. D*, **88**, 023516
[10] Jolien D. E. Creighton and Warren G. Anderson 2011 *Gravtitaional-Wave Physics and Astronomy* (New York: Wiley-VCH)
[11] E. Thrane, N. Christensen, and R. M. S. Schofield, 2013 *Phys. Rev. D*, **87**, 123009
[12] W. O. Schumann, 1952 *Zeitschrift fur Naturforschung A*, **7**, 149
[13] W. O. Schumann, 1952 *Zeitschrift fur Naturforschung A*, **7**, 250

Neutron Star Mass Distribution in Binaries

Chang-Hwan Lee and Young-Min Kim

Department of Physics, Pusan National University, Busan 46214, Korea

E-mail: clee@pusan.ac.kr, ymkim715@pusan.ac.kr

Abstract. Massive neutron stars with $\sim 2M_{\odot}$ have been observed in neutron star-white dwarf binaries. On the other hand, well-measured neutron star masses in double-neutron-star binaries are still consistent with the limit of $1.5M_{\odot}$. These observations raised questions on the neutron star equations of state and the neutron star binary evolution processes. In this presentation, a hypothesis of super-Eddington accretion and its implications are discussed. We argue that a $2M_{\odot}$ neutron star is an outcome of the super-Eddington accretion during the evolution of neutron star-white dwarf binary progenitors. We also suggest the possibility of the existence of new type of neutron star binary which consists of a typical neutron star and a massive compact companion (high-mass neutron star or black hole) with $M \geq 2M_{\odot}$.

1. Neutron Star Mass Observations

Neutron stars provide opportunities to investigate physics of low-temperature dense Quantum chromodynamics (QCD) matter and are sources of gravitational waves that can be detected by ground-based observatories such as LIGO[1], Virgo[2, 3] and KAGRA[4].

The central densities of neutron stars are expected to reach several times the normal nuclear matter saturation density (2.04×10^{14} g cm^{-3}). Many theoretical investigations provide mass-radius relations of neutron stars based on the predicted neutron star equations of state. However, due to the nature of strong interaction which governs the QCD matter, the inner structure of neutron stars are still unknown and the maximum mass of neutron star is still an open question [5]. Until 2010, no firm evidence of a high-mass ($\geq 2M_{\odot}$) neutron star has been found. High-mass neutron stars were listed in the X-ray binary catalogue, but they were not taken as the evidences of the existence of high-mass neutron stars due to the large uncertainties. The main uncertainty in mass estimates of X-ray binaries lies in finding the actual gravitational center of companion stars. The situation has been changed due to discoveries of two $\sim 2M_{\odot}$ neutron stars in neutron star-white dwarf binaries [6, 7]. These observations rule out many neutron star equations of state which predict the maximum mass of a neutron star to be less than $2M_{\odot}$. Recently, we investigated many equations of state that predict a larger neutron star maximum mass than $2M_{\odot}$ [8, 9, 10].

Double-neutron-star binaries are major sources for the neutron star mass measurements and gravitational-wave detections with current ground-based gravitational-wave observatories [11, 12]. All well-measured neutron star masses in double-neutron-star binaries are still less than $1.5M_{\odot}$, contrary to the recent $2M_{\odot}$ observation in neutron star-white dwarf binaries. These observations indicate that a neutron star's mass may depend on the evolutionary processes. In this presentation, we discuss a hypothesis of super-Eddington accretion during the evolution of binary progenitors.

2. Evolution of Neutron Star Binaries

Lee et al. [13] worked out the evolution of soft X-ray transient sources, i.e., black holes in binaries with main sequence or more-evolved companions. Based on the common envelope evolution, they estimated masses and spins of black holes at the time of black hole formation and found a correlation between the black hole's mass and binary's orbital period. Their predictions on black hole spins were confirmed by Shafee et al. [14] and McClintock et al. [15]. Moreno Méndez et al. [16] further discussed the consequences of black hole spins in soft X-ray transient sources to gamma-ray bursts and hypernovae. Based on the success of the evolution of black hole binaries, Lee and Cho [17] applied the same common envelope evolution [13] to the evolution of neutron star binaries. The key ingredient in their scenario is the super-Eddington accretion during the common envelope evolution.

In a spherical accretion, the Eddington luminosity is the maximum luminosity at which radiation pressure balances with the gravitational force. The main source of an accreting star's luminosity is the gravitational energy of accreting material. The Eddington accretion rate is the mass accretion rate with 100% conversion efficiency from the gravitational energy to the luminosity. For a typical accreting star, the Eddington accretion rate is considered to be the upper limit of the mass accretion rate. However, in the case of a core collapsed supernova, a neutron star is formed in a time scale of minutes with hyper-Eddington accretion rate, reaching 10^8 times of the Eddington accretion rate. If the Eddington accretion limit is hold, neutron stars cannot be formed from core collapsed supernovae. The hypothesis of super-Eddington accretion with an accretion rate of $10 \sim 100$ times of the standard Eddington accretion rate, have been discussed in the literature [18] and it is considered to be the origin of some of ultraluminous X-ray sources [19]. In this work, we assume that the super-Eddington accretion is hold during the common envelope phase in which the primary (first-born) neutron star spirals into the expanding envelope of a companion star [17, 20].

In order to understand the distribution of neutron star mass for a binary population, one has to take into account the evolution of neutron star binary progenitors [17, 20]. Since we are interested in neutron star binaries, we only consider a case in which the mass of more massive progenitor in a binary is large enough to produce a neutron star at the end of its evolution. Depending on the mass of the less massive progenitor, the secondary becomes either a neutron star or a white dwarf. The key property in this evolution is the initial mass difference ΔM between two progenitors in a binary. In this work, we assume that two progenitors are formed more or less at the same time in a given galaxy. One can consider a few simple cases for the comparison of different evolution processes.

- Case 1: $\Delta M \ll 4\%$
  When the mass difference is very small, two progenitors go into H-shell and He-shell burning at the same time such that two neutron stars are formed almost simultaneously and there is no time for the extra accretion. The Fe core masses of evolving giants at the final stage determine the final masses of neutron stars. Brown et al. [21] argued that the Fe core masses of giants in close binaries never grow beyond $1.5M_\odot$ because most of the H envelope is expelled during the common envelope phase and the giant evolves without H envelope at the later stage. The reduction in the gravitational force due to the absence of H envelope results in the less massive Fe cores compared to single star evolution. In this case, the final masses of neutron stars are less than $1.5M_\odot$ mainly because of the Fe core masses, independently of neutron star equations of state.
- Case 2: $\Delta M \leq 4\%$
  If the mass difference is not negligible even though $\Delta M \leq 4\%$, the secondary companion is already in H-shell burning stage when the primary neutron star is formed [17]. There is no time for the accretion during H-shell burning stage and the neutron star can accrete

Journal of Physics: Conference Series **716** (2016) 012021 doi:10.1088/1742-6596/716/1/012021

only at the He-shell burning stage . In this case, the primary neutron star can accrete up to $0.2M_\odot$ only during the He-shell burning stage .

- Case 3: $\Delta M > 4\%$
  When the mass difference is large enough, the more massive giant produces a neutron star while the less massive companion is still in main sequence. Even in this case, the Fe core mass of more massive giant in a close binary is still less than $1.5M_\odot$ because the H envelope is expelled during the common envelope phase. Hence the mass of the primary (first-born) neutron star is still less than $1.5M_\odot$ at birth. Now, the companion evolves into giant after the formation of the primary neutron star. In close binaries, the primary neutron star goes into the expanding envelope of the companion and has chances for the extra accretion after its formation. Lee and Cho [17] estimated that the primary neutron star can accrete up to $\sim 1M_\odot$. When the evolution is finished, the companion can also produce a compact star, i.e., a white dwarf or a neutron star.

The hypothesis of super-Eddington accretion depends on many other physical parameters such as spin of the progenitor, composition of the giant, etc. In this presentation, we mainly focus on the mass difference.

3. Mass Distribution in Neutron Star Binaries

The distribution of final masses of primary neutron stars in neutron star binaries are summarised in Fig. 1. In this figure, case 1 and 2 are indicated by a blue line and an arrows (NS-NS binaries) which are consistent with current observations ($\leq 1.5M_\odot$) of neutron star masses in double-neutron-star binaries. Case 3 is indicated by blue hatched area in the figure. Usually, the progenitor mass difference is larger than 4% for NS-WD binaries and the primary neutron star can accrete significant amount of material, reaching to $2M_\odot$. Since the masses of the progenitors of white dwarfs are less than $\sim 8M_\odot$. Observation of $2M_\odot$ neutron stars in NS-WD binaries is consistent with our estimation in Fig. 1.

The new feature in our approach is the possibility of a new class of neutron star binaries. For the binaries corresponding to the upper half part of Fig. 1, the final mass of the first-born neutron star is larger than $2M_\odot$ while the companion also produce a secondary neutron star. Depending on the neutron star equation of state, the first-born neutron star remains as a high-mass ($> 2M_\odot$) neutron star (HMNS) or collapses into a black hole.

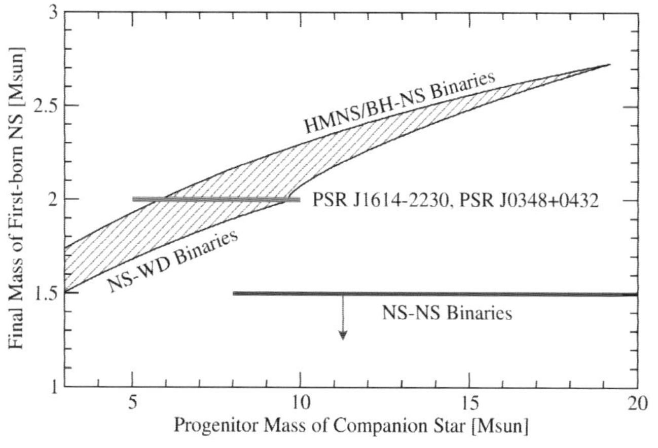

Figure 1. Final masses of primary (first-born) neutron stars in neutron star binaries as a function of the progenitor mass of the companion. Note that $2M_\odot$ neutron stars are observed only in neutron star-white dwarf (NS-WD) binaries [6, 7] and well-measured masses in double-neutron-star (NS-NS) binaries are still less than $1.5M_\odot$ [5] as indicated in the figure. This figure is taken from Lee and Cho [17].

4. Implication to Gravitational Waves

There is no observational evidence on the existence of new class of neutron star binaries. If the maximum mass of a neutron star is close to $\sim 2M_\odot$, the primary neutron star in the new class collapses into a black hole. The chance of observing black hole-neutron star binaries as pulsar binaries is very low mainly because there is no recycled pulsar since the first-born neutron stars collapsed into black holes. If the maximum mass of neutron star is much higher than $2M_\odot$, e.g. $2.5M_\odot$, the primary neutron star will remain as HMNS even with super-Eddington accretion. The probability of observing HMNS-NS binaries as pulsar binaries is not easy to predict because it depends on many other physical parameter (e.g., population, magnetic field strength, spin, inner structure, beaming angle, etc.) which we didn't include in the analysis. However, independently of whether the primary neutron stars remain as neutron stars or evolve into black holes, they are important sources of gravitational waves that can be detected by ground-based gravitational-wave observatories. Since the chirp mass depends on the total mass, in principle, one may be able to distinguish HMNS/BH-NS binaries from typical NS-NS binaries by detecting gravitational waves from these sources [22, 23].

Acknowledgements

This work was supported by a 2-Year Research Grant of Pusan National University.

References

[1] Harry G M (tor the LIGO Scientific Collaboration) 2010 *Class. Quantum Grav.* **27** 084006
[2] Acernese F et al. (The Virgo Collaboration) 2009 *Virgo Technical Report* **VIR-0027A-09** https://tds.ego-gw.it/ql/?c=6589
[3] Accadia T et al. (The Virgo Collaboration) 2012 *Virgo Document* **VIR-0128A-12** https://tds.ego-gw.it/ql/?c=8940
[4] Somiya K (for the KAGRA Collaboration) 2012 *Class. Quantum Grav.* **29** 124007
[5] Prakash M 2013 Neutron Stars and the EOS *Proceedings of 8th International Workshop on Critical Point and Onset of Deconfinement (to appear in Proceedings of Science) Preprint arXiv:1307.0397*
[6] Demorest P B, Pennucci T, Ransom S M, Roberts M S E, & Hessels J W T 2010 *Nature* **467** 1081
[7] Antoniadis J et al. 2013 *Science* **340** 448
[8] Lim Y, Kwak K, Hyun C H and Lee C-H 2014 *Phys. Rev.* D **89** 055804
[9] Lim Y, Hyun C H, Kwak K and Lee C-H 2014 ΛΛ interaction and neutron stars *Preprint arXiv:1412.5722*
[10] Lim Y, Hyun C H and Lee C-H 2015 Nuclear equation of state and neutron star cooling *Preprint arXiv:1501.04397*
[11] Abadie J et al (LIGO Scientific Collaboration and Virgo Collaboration) 2010 *Class. Quantum. Grav.* **27** 173001
[12] Dominik M et al. 2015 *Astrophys. J.* **806** 263
[13] Lee C-H, Wijers R A M J and Brown G E 2002 *Astrophys. J.* **575** 996
[14] Shafee R, McClintock J E, Narayan R, Davis S W, Li L and Remillard R A 2006 *Astrophys. J.* **636** L113
[15] McClintock J E, Shafee R, Narayan R, Remillard R A, Davis S W and Li L-X 2006 *Astrophys. J.* **652** 518
[16] Moreno Méndez E, Brown G E, Lee C-H and Walter F M 2011 *Astrophys. J.* **727** 29
[17] Lee C-H and Cho H S 2014 *Nuclear Physics* A **928** 296.
[18] Brown G E, Lee C-H and Bethe H A 2000 *Astrophys. J.* **541** 918
[19] Lee C-H and Brown G E 2003 *International Journal of Modern Physics* A **18** 527
[20] Lee C-H, Park H-J and Brown G E 2007 *Astrophys. J.* **670** 741
[21] Brown G E, Heger A, Langer N, Lee C-H, Wellstein S and Bethe H A 2001 *New Astronomy* **6** 457
[22] Allen B, Anderson W G, Brady P R, Brown D A, and Creighton J D E 2012 *Phys. Rev. D* **85** 122006
[23] Abadie J et al. (LIGO Scientific Collaboration and Virgo Collaboration) 2012 *Phys. Rev. D* **85** 082002

Active damping performance of the KAGRA seismic attenuation system prototype

Yoshinori Fujii[1], Takanori Sekiguchi[2], Ryutaro Takahashi[3], Yoichi Aso[3], Mark Barton[3], Fabián Erasmo Peña Arellano[3], Ayaka Shoda[3], Tomotada Akutsu[3], Osamu Miyakawa[2], Masahiro Kamiizumi[2], Hideharu Ishizaki[3], Daisuke Tatsumi[3], Naoatsu Hirata[3], Kazuhiro Hayama[4], Koki Okutomi[3], Takahiro Miyamoto[2], Hideki Ishizuka[2], Riccardo DeSalvo[5] and Raffaele Flaminio[3]

[1]Department of Astronomy, University of Tokyo, Bunkyo, Tokyo 113-0033, Japan
[2]Institute for Cosmic Ray Research, Kashiwanoha, Kashiwa-shi, Chiba 277-8582, Japan
[3]National Astronomical Obsevatory of Japan, Osawa, Mitaka-shi, Tokyo 181-8588, Japan
[4]Osaka City University, Sumiyoshi-ku Sugimoto, Osaka-shi, Osaka 558-8585, Japan
[5]University of Sannio, Corso Garibaldi 107, Benevento 82100, Italy

E-mail: yoshinori.fujii@nao.ac.jp

Abstract. The Large-scale Cryogenic Gravitational wave Telescope (formerly LCGT now KAGRA) is presently under construction in Japan. This May we assembled a prototype of the seismic attenuation system (SAS) for the beam splitter and the signal recycling mirrors of KAGRA, which we call Type-B SAS, and evaluated its performance at NAOJ (Mitaka, Toyko). We investigated its frequency response, active damping performance, vibration isolation performance and long-term stability both in and out of vacuum. From the frequency response test and the active damping performance test, we confirmed that the SAS worked as we designed and that all mechanical resonances which could disturb lock acquisition and observation are damped within 1 minute, which is required for KAGRA, by the active controls.

1. Introduction

KAGRA is a large cryogenic laser interferometric gravitational wave (GW) detector which is being constructed underground at the Kamioka mine in Gifu Prefecture Japan [1]. To avoid the GW signals from being buried in seismic noise, the mirrors of the interferometer are suspended by multi-stage pendulums called the seismic attenuation system (SAS), to prevent seismic motion from transferring to the mirrors and to let the mirrors move as free test masses. By suspending the mirrors, we can attenuate the seismic noise at frequencies higher than the resonance frequency of the SAS. However, the test mass would vibrate with large amplitudes at its mechanical resonance frequencies, so in order to lock the interferometer, we have to damp these resonances.

In KAGRA, we use three types of SAS for different tasks: Types A, B and Bp. The Type A SAS is an 8-stage pendulum system for the test masses (in the final full configuration which will be used for observation). For the beam splitter and signal recycling mirrors, we will use a 5-stage pendulum system called Type B SAS. The Type Bp SAS is a 3-stage pendulum system for the power recycling mirrors. In this paper, we report the Type B prototype test. In section

2, we summarize the design of the Type B SAS and its control strategy, and in section 3 the results are described.

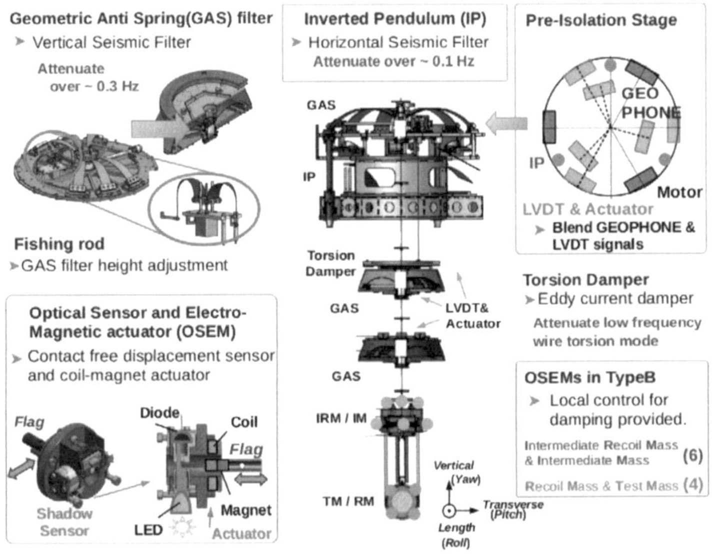

Figure 1. Overview of the Type B SAS for the beam splitter and signal recycling mirrors. The yellow boxes in the center show the position of the LVDTs and the actuators, while the red and green dots show the position of the OSEMs. We can actuate the top stage in 3 DoF, each GAS filter in the vertical direction, the IM for all 6 DoF, and the TM in 3 DoF.

2. Type B SAS

2.1. Mechanical details

The mechanical components of the Type B SAS are as follows (see also Figure 1);

- Inverted Pendulum (IP): three inverted pendulum legs support the whole system, and work as a horizontal seismic filter whose resonance frequency can be tuned as low as 0.1 Hz. The IP can attenuate not only the micro seismic noise at around 0.2 Hz, but also damp the multi-stage pendulum resonances around 1 Hz.
- Geometric Anti-Spring (GAS) filters: each filter is a vertical seismic isolator whose resonance frequency is tuned around 0.3 Hz. Type B SAS has three GAS filters. Each GAS filter consists of a set of curved maraging steel blades. It also has a spring height adjustment, in order to compensate thermal drift, attained through a motorised blade with low stiffness set in parallel to the GAS springs, called a fishing rod. The fishing rod can compensate weight of a few hundred grams.
- Torsion Damper: a non-contacting damper is placed above the middle GAS filter, which aims for damping low-frequency wire torsion modes of the SAS using eddy current damping.
- Payload: a 2-stage, 4-mass pendulum, which consists of the intermediate recoil mass (IRM), intermediate mass (IM), recoil mass (RM) and test mass (TM). We can adjust tilt and longitudinal displacement of the TM and the RM by applying forces on the TM from the RM.

Each component has sensors and actuators: The IP has three sets of coil-magnet actuators and two kinds of sensors arranged in a pinwheel configuration. Linear variable differential transformers (LVDTs) are used at low frequencies and geophones are used at high frequencies, with a crossover around 50 mHz. Each GAS filter has an LVDT as a vertical displacement sensor and a coil-magnet actuator. The payload is equipped with Optical Sensors and Electro-Magnetic actuators (OSEMs) [2] (see Figure 1), which combine shadow sensors and coil-magnet actuators. The linear range of the OSEMs is around 1 mm. By using 6 OSEMs on the IM and IRM, and 4 OSEMs on RM and TM, we can sense and actuate the relative motion of the IRM and IM in 6

DoF, and the RM and TM in 3 DoF. The signals from those sensors are to damp the resonances. An optical lever is also set at the front side of the TM to align the TM angular displacement.

2.2. Local active control strategy
Active damping controls are required to suppress the motion caused by the external disturbance driving the mechanical resonances of the SAS. The control system is set up to progress through the following states of operation: (a) damping, (b) lock acquisition and (c) observation [2]. The purpose of the damping state is purely to reduce large-amplitude motion at the mechanical resonances due to external disturbances. In this state the active controls are required to minimize the decay time of the SAS to allow fast recovery to the observation state. We have set a goal for the required damping time of 1 min. In the lock acquisition state the speed of the mirrors has to be suppressed to allow the mirrors to be trapped into the linear range of the interferometer signals. The control noise must be designed not to increase the RMS of linear speed of each suspended mirror. To lock the optical cavities, the RMS velocities and the angular fluctuations of the mirrors are typically required to be less than 1 μm/sec and 1 μrad respectively [3]. After lock acquisition is achieved, we enter the observation state, in which the interferometer operates. In each state, we use different control system, for example, OSEMs are used only in the damping and lock acquisition states because of their large noise coupling. Conversely, geophones are used in the lock acquisition ans observation state due to their small range.

3. Performance test of type B SAS prototype

Using the Type B SAS prototype, we constructed a servo system for damping (see Figure 2) and conducted performance tests of the prototype: frequency response, active damping performance, vibration isolation performance and long term stability performance. In this section the measured frequency response and active damping performance are summarized.

3.1. Frequency response
To confirm the SAS works as designed we measured transfer functions from actuators to sensors, and compared them with results from a rigid-body simulation [4]. The transfer functions were measured for all the 15 degrees of freedom (DoFs) that we can actuate, in air and with the controls switched off. A typical result is shown in Figure 3, where the blue and red lines show the simulated and measured results. The measured curves fitted well the predictions.

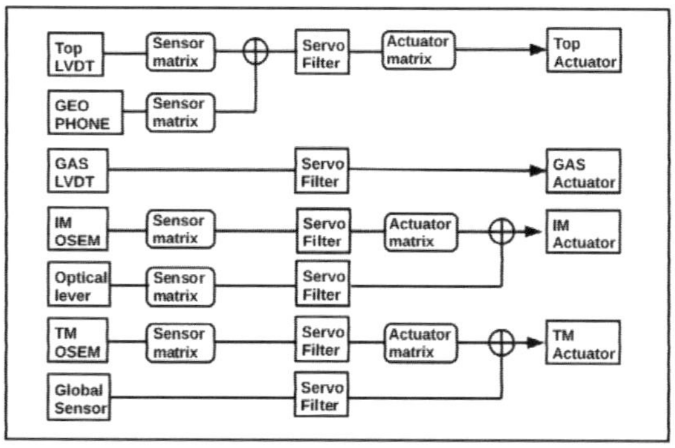

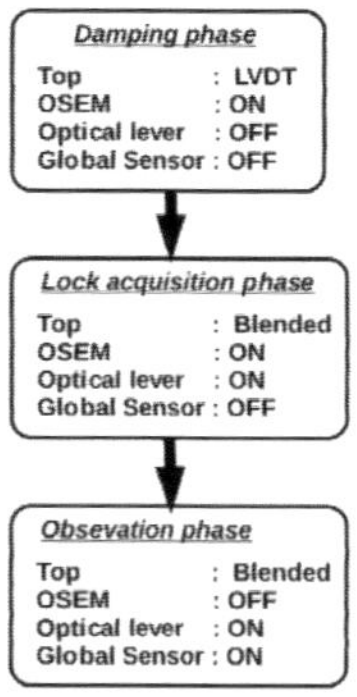

Figure 2. Left: local damping control for performance test of the type B SAS prototype. Signals from photo sensors are used instead of as-yet unavailable main interferometer signals. Right: local control transition toward observation state.

3.2. Active damping performance in damping state
To test if the mechanical resonances of the SAS were damped within 1 min, which is required for KAGRA, by the active controls in the damping state, we measured the $1/e$ damping time with the controls on and off. In this measurement, we obtained the damping time of the mechanical resonances by exciting the system using appropriate actuators and measuring decay signals using

implemented sensors. Also, only modes with resonance frequencies below 20 Hz were measured because of the difficulty in exciting the system above 20 Hz with actuators on the SAS. The results is shown in Figure 4. We found basically the resonances were damped properly. Although a decay time of one mode was slow, we confirmed we could damp within 1 min all the resonances which were likely to disturb lock acquisition and observation. For only one resonance at 0.6 Hz we could not reduce the decay time below 1 min, however, this resonance is hard to excite other than by external disturbance. The mode is concerning with transversal motions of TM and RM. Even if it becomes excited, it cannot affect the lock acquisition unless the amplitude is extremely large, such as 1 mm, which is the requirement of the beam spot fluctuation on the optics of the interferometer.

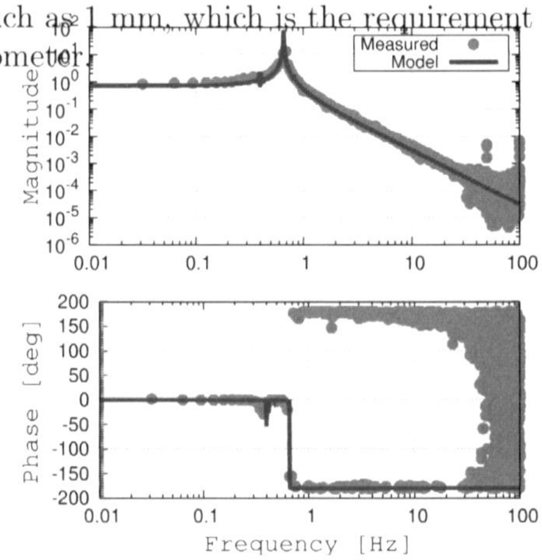

Figure 3. Transfer function of the TM Length DoF in the air

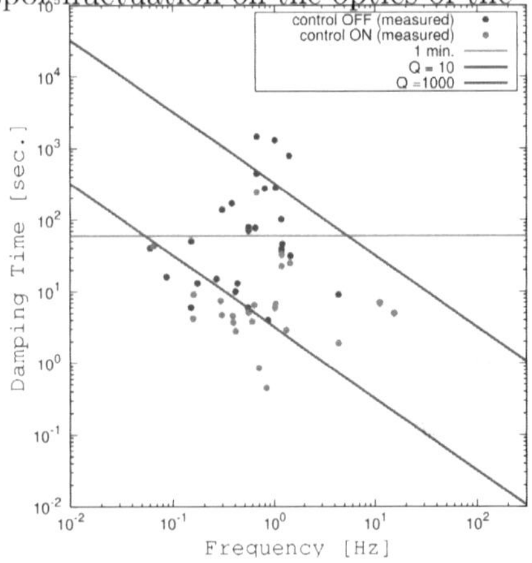

Figure 4. Active damping performance of the prototype

4. Summary

We assembled a Type B SAS prototype and evaluated its performance. From the frequency response test, we confirmed all the components of the SAS were properly suspended. From the active damping performance test, we confirmed we were able to damp the mechanical resonances caused by external disturbances and recover the steady state again within a few minutes.

Acknowledgments

This work was supported by MEXT, JSPS Leading-edge Research Infrastructure Program, JSPS Grant-in-Aid for Specially Promoted Research 26000005, MEXT Grant-in-Aid for Scientific Research on Innovative Areas 24103005, JSPS Core-to-Core Program, A. Advanced Research Networks, and the joint research program of the Institute for Cosmic Ray Research, University of Tokyo, and JSPS Grant-in-Aid for JSPS Fellows 15J11064, 14J10236. We thank personnel of the Advanced Technology Center of the National Astronomical Observatory of Japan for providing items of equipment for our experiments.

References

[1] Aso Y, et al. 2013, Physics Review D 88 043007
[2] Sekiguchi T, 2016, Ph.D thesis
[3] Michimura Y, 2012, JGW document JGW-T1202403-v1
[4] Sekiguchi T, 2012, Master's thesis

Thermal connection and vibrational isolation: an elegant solution for two problems

C Frajuca[1*], F S Bortoli[1], N S Magalhaes[2] and O D Aguiar[3]

[1] Sao Paulo Federal Institute, Rua Pedro Vicente 625, Sao Paulo, SP, 01109-010, Brazil
[2] Federal University of Sao Paulo, Department of Exact and Earth Sciences, Rua Sao Nicolau 120, Diadema, SP, 09913-030, Brazil
[3] INPE Astrophysics Division, Sao Jose dos Campos, SP, 12227-010, Brazil

E-mail: frajuca@gmail.com

Abstract. Schenberg is a detector of gravitational waves resonant mass type, with a central frequency of operation of 3200 Hz. Transducers located on the surface of the resonating sphere, according to a distribution half-dodecahedron, are used to monitor a strain amplitude. To improve the performance of the detector it is essential to decrease the temperature, then it will be cooled down, this temperature could reach as low as 50 mK. This refrigerator produces vibration noise that could compromise the performance of Schenberg detector. In this work we the study such vibration noise and how it could be minimized proposing a new connection from the dilution refrigerator to the sphere suspension. The vibration attenuation is studied by finite element modeling (FEM) and an attenuation higher than 10^{24} is found, higher enough to note compromise the performance of Schenberg detector.

1. Introduction

SCHENBERG is a spherical resonant-mass Gravitational Wave (GW) detector [1, 2] built in Sao Paulo, Brazil, can be seen on figure 1. The sphere with 65 cm in diameter and weighting 1.15 ton, is made of a copper-aluminum alloy [3] with 94 % Cu and 6 % Al. The detector will have six electromechanical transducers (motion sensors that monitor the sphere surface movement, transforming the mechanical oscillation in electrical signal [4, 5]), arranged on the spheres surface in a half-dodecahedron distribution. The sensors will be located as if in the center of the six connected pentagons in a dodecahedron surface, following the studies by Merkowitz and Johnson [6, 7] confirmed by Magalhaes and collaborators [8]. By analyzing the signal of such sensors, the amplitudes and the direction of the incoming gravitation wave can be determined [9, 10]. The history of the decisions process that has determined Schenberg characteristics can be seen in [12]. The sphere, in a commissioning phase is cooled to 4.2 Kelvin. In a next phase will be cooled down to a lower temperature using a dilution refrigerator, this temperature could reach as low as 50 mK. This refrigerator produces noise because of the Helium evaporation and this noise is transported by the connections to the sphere. In this work we the study such vibration noise and how it could be minimized. The conventional method used in detectors with this kind of refrigerator is to connect the refrigerator to the sphere using thin copper wires, but it reduces the cooling capability by a great factor. The vibration attenuation should make the dilution refrigerator noise lower than the thermal vibration noise on the sphere surface keeping the temperature as lower as possible. Figure 1 shows Schenberg schematics. The analysis

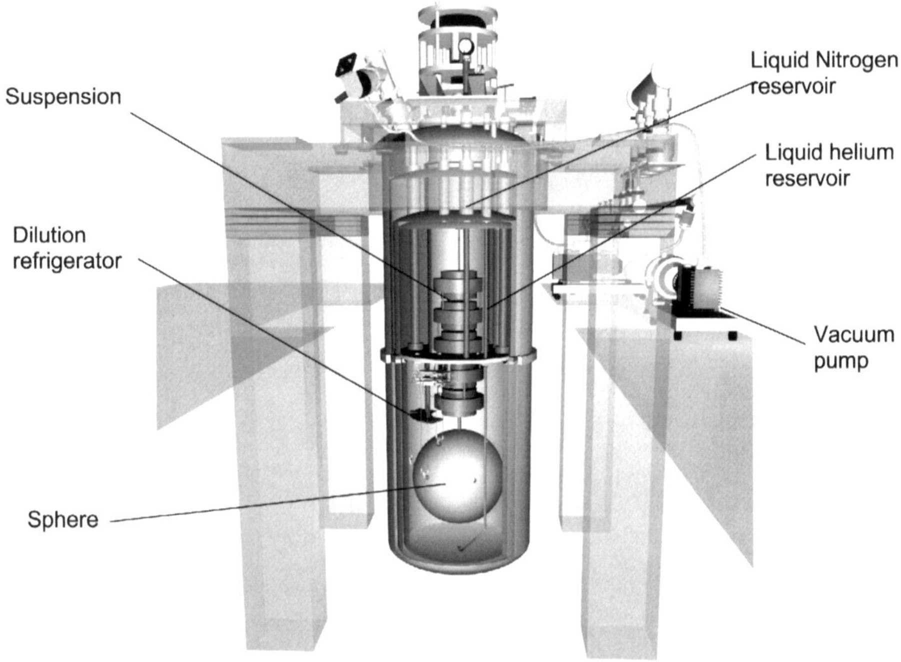

Figure 1. Schenberg schematics, the green area is liquid nitrogen and the blue one is filled with liquid helium.

of the vibration attenuation is done by frequency response obtained finite element modeling (SolidWorks Simulation 2010-2011). Figure 2 shows the the 1 K pot in Schenberg detector and the mixing chamber.

2. Requirement for the noise attenuation

In order to allows the Schenberg detector to measure vibrations due to gravitational waves instead of vibrations due to diluiton refrigerator noise, it is essential to attenuate such noise to an amplitude below the noise to the thermal vibrations. The Noise vibration on a sphere surface due to thermal noise at a temperature of 50 mK is equal to 10^{-46} m^2/Hz [7]. An estimation for the dilution refrigerator noise can be find in [?] as a value below 10^{-22} m^2/Hz. Then, in order to get the noise below the thermal noise, an attenuation of, at least, 10^{24} is needed.

3. The Mixing Chamber connected to the last but one suspension mass

Using the idea that a mass connected to a wire can reflects part of the vibration traveling across the wire, a series of masses connected to parallel wires should do the same. Figure 3 shows the proposed connection from the dilution refrigerator to the sphere suspension.

4. Frequency response from a 10^{-2} m^2/Hz signal on the mixing chamber disk

In order to estimate the attenuation of the proposed connection a finite element modeling was performed using SolidWorks Simulation 2010-2011 version, with a dumping ratio of 0.0005. Figure 4 shows the frequency response obtained on the sphere surface due to a white noise of

Figure 2. Schenberg suspension with the 1K pot dilution refrigerator and, on the right, the mixing chamber.

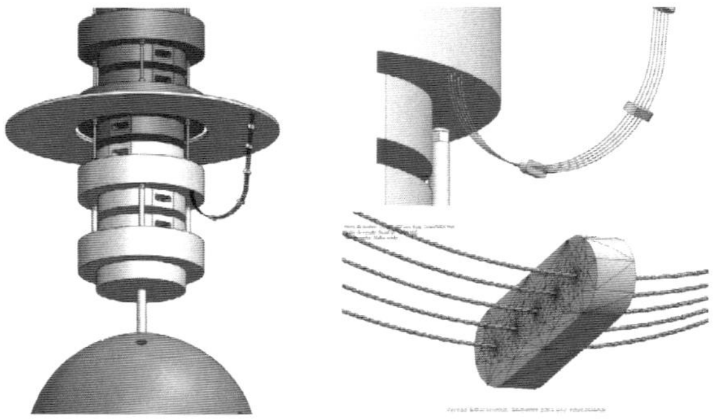

Figure 3. The proposed connection, the wires could be welded to the masses.

the 10−2 m^2/Hz signal applied on the mixing chamber disk, the value of such displacement is chosen because of the limitation on the finite elements program.

5. Conclusions

We considered the vibration noise on the sphere surface as 10^{-46} m^2/Hz due to thermal noise and the vibration noise in the region close to the dilution refrigerator was estimated as 10^{-22} m^2/Hz. One configuration of the connection of the mixing chamber were modeled in a finite element program. By the results of this finite element modeling, the dilution refrigerator when connected to the suspension last but one mass, has an attenuation equal to 10^{24} found in the frequency response graphic. Barely the attenuation needed as the total vibration on the sphere surface due to the dilution refrigerator noise is of the order of the thermal noise. The noise considered should be much lower due to a concrete weighting of 4 ton, where the dewar is supported, such calculation is difficult as the size of the model becomes too big. Considering this 4 ton mass, the noise should be much lower than the one estimated above.

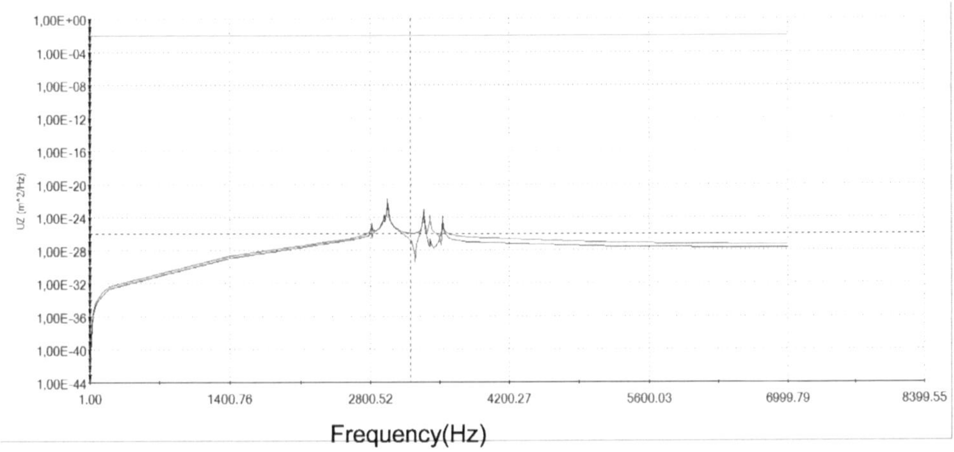

Figure 4. Frequency response on the sphere surface.

Acnowledgements
CF acknowledges FAPESP (grant # 2013/26258-4), CNPq (grant # 312906/2013-7) for the financial support.

References
[1] Aguiar O D et al 2012 *J. Phys.: Conf. Ser.*, **363**(1) 012003
[2] Frajuca C, Ribeiro K L, Aguiar O D, Andrade L A, Castro P J, Magalhaes N S and Marinho Jr R M 2004 *CQG* **21** 1107
[3] Frossati G 1997 *Proceedings of the First International Workshop for an Omnidirectional Gravitational Radiation Observatory*, Velloso, Jr. W F, Aguiar O D and Magalhaes N S, editors (Singapore, World Scientific)
[4] Frajuca C, Ribeiro K L, Andrade L A, Velloso Jr W F, Melo J L, Aguiar O D and Magalhaes N S 2002 *CQG* **19** 1961
[5] Frajuca C, Aguiar O D, Magalhaes N S, Ribeiro K L and Andrade L A 1999 *Proc. 3rd Edoardo Amaldi Conference on Gravitational Waves (Pasadena, USA, July 1999) and AIP Conf. Proc.* **523** (New York, AIP) p. 417.
[6] Johnson W W and Merkowitz S M 1993 *Phys. Rev. Lett.* **70** 2367
[7] Merkowitz S M and Johnson W W 1997 *Phys. Rev. D* **56** 7513
[8] Magalhaes S M, Aguiar O D, Johnson W W and Frajuca C 1997 *Gen. Relat. Grav.* **29** 1511
[9] Magalhaes N S, Johnson W W, Frajuca C and Aguiar O D 1995 *MNRAS* **274** 670
[10] Magalhaes N S, Jonhson W W, Frajuca C and Aguiar O D 1997 *Astrophysical Journal* **475** 462
[11] Aguiar O D et al 2002 *Braz. J. Phys.* **32** 4 p. 866 (Dec)
[12] Oleg K, Junyum L, Larry L, Alvin A and Vladimir M 2003 *Physica B* **329-333** 1604

Journal of Physics: Conference Series **716** (2016) 012024 doi:10.1088/1742-6596/716/1/012024

Gravitational radiation from a particle in bound orbit around a black hole; relativistic correction

Ashok Tiwari, Udayaraj Khanal

Central Department of Physics, Tribhuvan University, Kathmandu, Nepal

E-mail: `tiwarias@yahoo.com`, `khanalu@yahoo.com`

Abstract. Gravitational radiation from a system of two body, one as test particle and other as black hole (we assume, m_1 is mass of the test particle and m_2 is mass of black hole in bound orbits (orbital eccentricities $e < 1$) and $E^2 < 1$; E is the energy, is calculated with relativistic correction using the method of inertia tensor and multipole formalism. Plots of power versus eccentricity of the bound orbit of first kind are presented, and average total power radiated as a function of eccentricity is plotted according to inertia tensor method. According to multipole formalism the power radiated in gravitational waves from an bound orbit is given by enhancement factor g(n,e) times the function of other parameters is plotted. The calculations apply for arbitrary eccentricity of the relative orbit, assuming orbital velocities are small.

1. Introduction

Accelerating heavenly bodies produces a disturbance in space-time which travels with the speed of light, called as gravitational waves. It is nothing but the perturbations of the flat space-time [5]. One of the most promising astrophysical source of gravitational radiation is the binary star system [4]. This system consists of binary NS, binary BH, and NS-BH binary. The binary star inspiraling each other radiates GR and come closer to each other. Then after certain time coalsec. The time of coalescence depends directly on the eccentricity of the orbit. The power of emitted radiation depends on the mass of black hole and the mass of the particle. It also depends on the semi-major axis and the eccentricity of the orbit before coalescence. Here in this work we used two method to calculate the power generated by point particle in bound orbit around the black hole.

2. Methodology

I. *Inertia Tensor Method*: Let the masses m_1 and m_2 have coordinates $(d_1 \cos\chi\ , d_1 \sin\chi)$ and $(-d_2 \cos\chi\ , -d_2 \sin\chi)$ in the xy plane as shown in figure 1. The origin will be taken to be at the center of mass so that,

$$d_1 = \left(\frac{m_2}{m_1 + m_2}\right) d, \quad d_2 = \left(\frac{m_1}{m_1 + m_2}\right) d \tag{1}$$

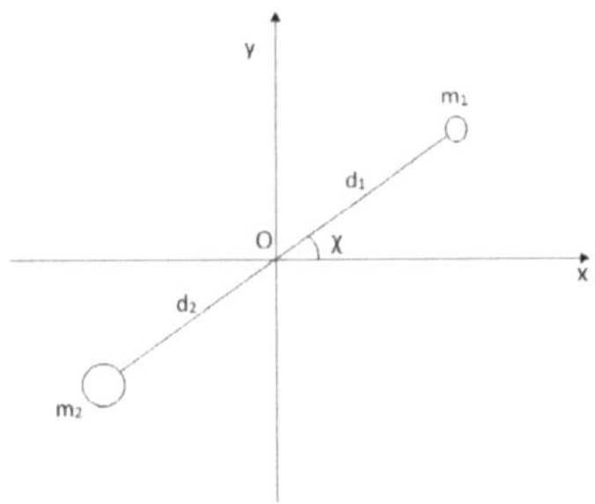

Figure 1. Coordinate system used in calculation.

Now the non-vanishing quadrupole tensor [3] are,

$$Q_{xx} = \mu d^2 \cos^2 \chi, \qquad Q_{yy} = \mu d^2 \sin^2 \chi, \qquad Q_{xy} = Q_{yx} = \mu d^2 \sin \chi \cos \chi \tag{2}$$

where, μ is the reduced mass of the binary system and the distance between two masses $d = d_1 + d_2$. From kepler motion, the orbit equation is,

$$u = \frac{1}{l}(1 + e \cos \chi) \tag{3}$$

where, $u = 1/d$, e is the eccentricity of the orbit and l is the latus rectum of the orbit.

We have equation with relativistic correction for bound orbit [1],

$$\pm \frac{d\chi}{d\phi} = (1 - 6\mu + 2\mu e)^{1/2} \left(1 - k^2 \cos^2 \chi/2\right)^{1/2} \tag{4}$$

where, k^2 depends on reduced mass of the system and eccentricity of the orbit and given by,

$$k^2 = \frac{4\mu e}{1 - 6\mu + 2\mu e}$$

Now, we can calculate the angular velocity ($\dot{\chi}$), with substitution of k^2, $\cos^2 \chi/2$ and $d\phi/dt$ is,

$$\frac{dt}{d\phi} = \frac{E}{\mathbf{L}u^2(1 - 2Mu)} \tag{5}$$

where, $\mathbf{L}$ denotes the angular momentum about an axis normal to the invariant plane and $M = m_1 + m_2$ i.e total mass of the binary system. Simplifying, we get angular velocity,

$$\dot{\chi} = \sqrt{1 - 6\mu - 2\mu e \cos \chi} \frac{\mathbf{L}(1 + e \cos \chi)^2}{El^2} \left[1 - \frac{2M}{l} - \frac{2Me \cos \chi}{l}\right] \tag{6}$$

We know that, $l = a(1 - e^2)$ where, a is the semimajor axis and e is the eccentricity of the ellipse(orbit). The total power is calculated from [3],

$$\begin{aligned} P \;=\; & \frac{G}{5c^5}[(\frac{d^3Q_{xx}}{dt^3}\frac{d^3Q_{xx}}{dt^3} + 2\frac{d^3Q_{xy}}{dt^3}\frac{d^3Q_{xy}}{dt^3} + \frac{d^3Q_{yy}}{dt^3}\frac{d^3Q_{yy}}{dt^3}) \\ & -\frac{1}{3}(\frac{d^3Q_{xx}}{dt^3}\frac{d^3Q_{xx}}{dt^3} + 2\frac{d^3Q_{xx}}{dt^3}\frac{d^3Q_{yy}}{dt^3} + \frac{d^3Q_{yy}}{dt^3}\frac{d^3Q_{yy}}{dt^3})] \end{aligned} \tag{7}$$

Where, G is gravitational constant = $6.67{*}10^{-11}N.m^2/kg^2$ and Q_{ij} are quadrupole tensors.

II. *Multipole Expansion*: In this method, we evaluate the m_{2M}, in terms of Q_{ij} they are [2],

$$m_{2\pm2} = \frac{ik\omega^3}{10\sqrt{3}}\left(\frac{15}{32\pi}\right)^{1/2}(Q_{xx} - Q_{yy} \pm 2iQ_{xy}) \tag{8}$$

$$m_{2\pm1} = 0 \tag{9}$$

$$m_{20} = \frac{-ik\omega^3}{10\sqrt{3}}\left(\frac{5}{16\pi}\right)^{1/2}(Q_{xx} + Q_{yy}) \tag{10}$$

Where, $k = \sqrt{32\pi G}$ and $\omega = n\omega_0$, $\omega_0 = [G(m_1 + m_2)/a^3]^{1/2}$ is the average angular velocity. Q_{ij} are calculated by using Fourier series of argument M with recursion relation of Bessel's function. Finally, the power radiated in the nth harmonics is,

$$P(n) = \frac{32G^4m_1^2m_2^2(m_1 + m_2)}{5c^5a^5}g(n, e) \tag{11}$$

Where,

$$\begin{aligned} g(n,e) \quad &= \quad \frac{n^4}{32}([J_{n-2}(ne) - 2eJn - 1(ne) + \frac{2}{n}J_n(ne) + 2eJ_{n+1}(ne) - J_{n+2}(ne)]^2 \\ &+(1-e^2)[J_{n-2}(ne) - 2J_n(ne) + J_{n+2}(ne)]^2 + \frac{4}{3n^2}[J_n(ne)]^2) \end{aligned} \tag{12}$$

3. Results

1. Inertia Tensor Method: It is straightforward to calculate the d^3Q_{ij}/dt^3: and we have,

$$\frac{d^3Q_{xx}}{dt^3} = \frac{d^3Q_{xx}}{d\chi^3}(\frac{d\chi}{dt})^3 + \frac{d^2Q_{xx}}{d\chi^2}\frac{d}{dt}(\frac{d\chi}{dt})^2 + \frac{d^2Q_{xx}}{d\chi^2}\frac{d\chi}{dt}\frac{d^2\chi}{dt^2} + \frac{dQ_{xx}}{d\chi}\frac{d^3\chi}{dt^3}$$

and similar results for other components.

We can calculate angular velocity ($\dot{\chi}$), using equation (6). Since, $l = a(1-e^2)$, taking semimajor axis (a) of an ellipse as constant, we vary eccentricity and after solving equation [6],

$$\frac{1}{L^2} = \frac{1}{lM}\left[1 - \mu(3 + e^2)\right]$$

We get,

$$L = l\sqrt{\frac{m_1m_2}{(m_1 + m_2) - 3m_1m_2 - m_1m_2e^2}}$$

We can calculate the latus rectum for ellipse, keeping semimajor axis as a constant. We can take $a = 11.458$, m_2 = mass of black hole = 1, m_1 = mass of test particle = 0.001

For bound orbit, $e < 1$ and $E^2 < 1$, to meet this requirement we put $E = 8/9$. The time period of the orbit, according to keplers law is,

$$T = \frac{2\pi a^{3/2}}{\sqrt{GM}}$$

Substituting $G = 1$ and $a = 11.458$ for simplicity, we get $1/T = 0.0041$.

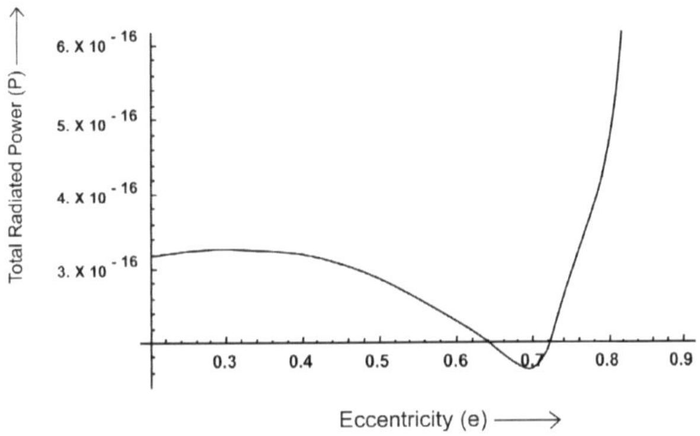

Figure 2. The relative power radiated with different eccentricities (e)

We calculated the numerical values of latus rectum and angular momentum($\mathbf{L}$) for different eccentricities. We put the numerical values for mass(M), eccentricity(e), latus rectum (l), angular momentum then we obtained total average radiated power. We take only real part of our calculation, because imaginary part is very-very small; at the order 10^{-48}. The total relative radiated power is plotted with eccentricity(e),which is shown in fig 2.

2. Multipole Expansion: From this method, we have calculated the total power radiated and we plot it with different $[g(n,e)]$: that is n^{th} harmonics, for eccentricities $e = 0.2, 0.5, 0.7$. We get the following plot; fig(3) using this method.

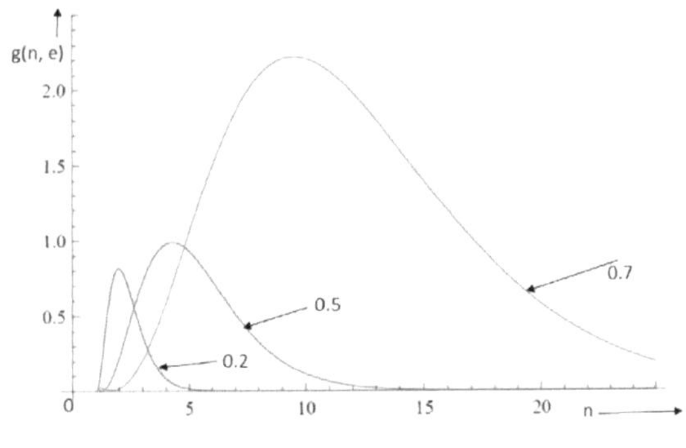

Figure 3. $[g(n,e)]$, the relative power radiated into the nth harmonics, for eccentricities $e = 0.2, 0.5, 0.7$

4. Conclusion

1. Inertia Tensor Method

We calculated the relative total power radiated with the application of relativistic correction in classical results. We found that the total radiated power is strongly depend on eccentricity of the orbit. With relativistic correction, we found that with increasing of eccentricity, total radiated power is also increasing, but on further increasing of eccentricity we found certain decrement in total power radiated, looks local minimum in figure 2. Again, on further increasing in eccentricity, the total relative power is steeply rising.This is from the relativistic correction in classical results. The nature of the plot is similar to the results of P.C. Peter's and J. Mathew's result's [3] except that the minimum doesn't appear in Newtonian result. We believe this minimum is due to relativistic correction. The radiation should depend so strongly on the eccentricity is not surprising. In electromagnetic radiation, the power radiated increases for increasing accelerations. Thus, the bodies will radiate most at their closest approach, and for fixed energy the higher the eccentricity, the higher the power radiated will be.

2. Multipole Expansion

From this method we calculate total radiated power in terms of the $g(n, e)$; the relative power radiated into the nth harmonic. The plot of $g(n, e)$ with different eccentricities $e = 0.2, 0.5, 0.7$ are shown in figure (3), with the increasing of harmonics (n) and eccentricity (e), we get large and smooth curve. This explains why the higher harmonics dominate the radiation for (e) near 1. Fourier components of large (n) must be present to give such a peaking of the radiation at one part of the path.

Acknowledgments

We would like to thank B.P Koirala memorial planeterium and observatory development board,*Ministry of Science and Technology, Government of Nepal* for providing partial support to perform this work.

References

[1] Chandrasekhar S., *The Mathematical Theory of Black Holes*, (Oxford University press, Walton Street, Oxford, 1983), chapter 3.

[2] Mathews *J., J. Soc. Ind. Appl. Math.* **10**, 768 (1962).

[3] Peters P. C. and Mathews J., 1963, *Phys. Rev.*, **131**, 435-440.

[4] Schutz B. F., *Gravitational Radiation*, (arXiv;gr-qc/0003069 v1 16 Mar 2000).

[5] Schutz B. F., *A First Course in General Relativity*, (Cambridge University press, 1985), chapter 9.

[6] Weinberg S.,*Gravitation and Cosmology: Principles and Applications of the General Theory of Relativity* (John Wiley and Sons Inc., U.K., 2008), chapter 10.

Newtonian noise cancellation in tensor gravitational wave detector

Ho Jung Paik[1] and Jan Harms[2]

[1] Department of Physics, University of Maryland, College Park, MD 20742, USA
[2] Università degli studi di Urbino "Carlo Bo" and INFN Sezione di Firenze, Sesto Fiorentino, I-50019, Italy

hpaik@umd.edu

Abstract. Terrestrial gravity noise produced by ambient seismic and infrasound fields poses one of the main sensitivity limitations in low-frequency ground-based gravitational-wave (GW) detectors. This noise needs to be suppressed by 3–5 orders of magnitude in the frequency band 10 mHz to 1 Hz, which is extremely challenging. We present a new approach that greatly facilitates cancellation of gravity noise in full-tensor GW detectors. It makes explicit use of the direction of propagation of a GW, and can therefore either be implemented in directional searches for GWs or in observations of known sources. We show that suppression of the Newtonian-noise foreground is greatly facilitated using the extra strain channels in full-tensor GW detectors. Only a modest number of auxiliary, high-sensitivity environmental sensors is required to achieve noise suppression by a few orders of magnitude.

1. Introduction

The Newtonian noise (NN) generated by moving local masses poses a formidable challenge to approaching the detector noise limit in gravitational-wave (GW) detectors at frequencies below 1 Hz. At low frequencies, the NN is dominated by Rayleigh waves and infrasound waves. For a laser interferometer to overcome this noise below 1 Hz, the ground and air motion within tens of kilometers from the detector must be measured with a large number of seismometers and microphones with sufficient accuracy, and then the induced NN computed and subtracted from the detector output. The NN from Rayleigh waves could be canceled up to one part in 10^3 by using this method [1]. But for infrasound waves, cancellation works only for waves coming *in certain favorable directions*.

In contrast, by using its tensor nature, SOGRO can mitigate the NN from both Rayleigh waves and infrasound waves to one part in 10^3 for *all incident angles*. A detailed analysis of NN mitigation for SOGRO has been published elsewhere [2]. In this paper, we summarize the result and discuss the possibility of using mini-SOGROs to mitigate NN for advanced laser interferometer GW detectors.

2. Mitigation of NN on SOGRO

Assuming that the interferometer is underground at depth $z < 0$, the gravitational perturbation of a single test mass due to a Rayleigh wave incident at an angle ψ with respect to the sensitive axis x of the test mass and an infrasound wave incident in direction (ψ, ϑ) is given [1] by

$$X(\omega) = -2\pi i \cos\psi G\rho_0\gamma_R \frac{\xi(\omega)}{\omega^2}\exp\left(\frac{\omega}{c_R}z\right) - 4\pi i G \sin\vartheta \cos\psi \frac{\delta\rho(\omega)c_{IS}}{\omega^3}\exp\left(\frac{\omega}{c_{IS}}z\sin\vartheta\right), \quad (1)$$

where $\xi(\omega)$ and $\delta\rho(\omega)$ are the vertical ground displacement and atmospheric density fluctuation directly above the test mass, $\gamma_R \approx 0.83$ is a factor that accounts for the partial cancellation for the Rayleigh NN from surface displacement by the sub-surface compressional wave content of the wave field, $c_R \approx 3.5$ km/s and $c_{IS} \approx 330$ m/s are the speed of the Rayleigh waves underground and the infrasound waves, respectively, and ρ_0 is the mean mass density of the ground.

The metric perturbation tensor *in the detector coordinates* can be shown to be

$$h_N(\omega) = \left[a(\omega)\xi(\omega) + b(\omega,\vartheta)\rho(\omega)\right]\begin{pmatrix} \cos^2\psi & \cos\psi\sin\psi & -i\cos\psi \\ \cos\psi\sin\psi & \sin^2\psi & -i\sin\psi \\ -i\cos\psi & -i\sin\psi & -1 \end{pmatrix}, \tag{2}$$

where

$$a(\omega) \equiv \frac{2\pi G}{\omega}\frac{\rho_0\gamma_R}{c_R}\exp\left(\frac{\omega}{c_R}z\right),\quad b(\omega,\vartheta) \equiv \frac{4\pi G}{\omega^2}\sin^2\vartheta\exp\left(\frac{\omega}{c_{IS}}z\sin\vartheta\right). \tag{3}$$

Consider a GW coming from (θ, ϕ) direction in the presence of *multiple* Rayleigh and infrasound waves. The full strain tensor *in the GW coordinates* has the form:

$$h'(\omega) = \begin{pmatrix} h_+(\omega) + h'_{N11}(\omega) & h_\times(\omega) + h'_{N12}(\omega) & h'_{N13}(\omega) \\ h_\times(\omega) + h'_{N12}(\omega) & -h_+(\omega) + h'_{N22}(\omega) & h'_{N23}(\omega) \\ h'_{N13}(\omega) & h'_{N23}(\omega) & h'_{N33}(\omega) \end{pmatrix}, \tag{4}$$

with

$$h'_{N11}(\omega) = \sum_i \left[a(\omega)\xi_i(\omega) + b(\omega,\vartheta_i)\rho_i(\omega)\right]\left[\cos(\psi_i - \phi)\cos\theta + i\sin\theta\right]^2, \tag{5a}$$

$$h'_{N22}(\omega) = \sum_i \left[a(\omega)\xi_i(\omega) + b(\omega,\vartheta_i)\rho_i(\omega)\right]\sin^2(\psi_i - \phi), \tag{5b}$$

$$h'_{N33}(\omega) = \sum_i \left[a(\omega)\xi_i(\omega) + b(\omega,\vartheta_i)\rho_i(\omega)\right]\left[\cos(\psi_i - \phi)\sin\theta - i\cos\theta\right]^2, \tag{5c}$$

$$h'_{N12}(\omega) = \sum_i \left[a(\omega)\xi_i(\omega) + b(\omega,\vartheta_i)\rho_i(\omega)\right]\sin(\psi_i - \phi)\left[\cos(\psi_i - \phi)\cos\theta + i\sin\theta\right], \tag{5d}$$

$$h'_{N23}(\omega) = \sum_i \left[a(\omega)\xi_i(\omega) + b(\omega,\vartheta_i)\rho_i(\omega)\right]\sin(\psi_i - \phi)\left[\cos(\psi_i - \phi)\sin\theta - i\cos\theta\right], \tag{5e}$$

$$h'_{N13}(\omega) = \sum_i \left[a(\omega)\xi_i(\omega) + b(\omega,\vartheta_i)\rho_i(\omega)\right]\left[\cos(\psi_i - \phi)\cos\theta + i\sin\theta\right]\left[\cos(\psi_i - \phi)\sin\theta - i\cos\theta\right]. \tag{5f}$$

Due to the *transverse nature* of the GW, h'_{13}, h'_{23} and h'_{33} contain only the NN components. To recover $h_+(\omega)$ and $h_\times(\omega)$, the NN could be removed from h'_{11} and h'_{12} by correlating them with h'_{13}, h'_{23} and h'_{33}, and possibly also with some CM channels and subtracting the correlated parts.

Figure 1 shows the residual NN achieved for Rayleigh waves *in the absence of* infrasound waves by using h'_{13}, h'_{23}, h'_{33} and a_z, plus seven seismometers with signal-to-noise ratio (SNR) of 10^3 at the radius of 5 km as the input of the Wiener filter. The NN has been removed to about 10^{-3} with environmental sensors (seismometers) alone. The local channels of SOGRO improve the noise significantly only near $\theta = \pi/2$, where the noise of the DM and CM channels drop out.

Figure 2 is the residual NN achieved for infrasound waves *in the absence of* Rayleigh waves by using h'_{13}, h'_{23}, h'_{33} and 15 microphones of SNR of 10^4, one at the detector, seven each at the radius of 600 m and 1 km around the detector. With the environmental sensors (microphones) alone, the NN *cannot* be mitigated except at $\theta = 0$, $\pi/2$ and π. This is because the infrasound waves come from a half space and the microphones deployed over a surface is insufficient to measure the effect of 3D density variations of the atmosphere. Thus mitigation of infrasound NN constitutes a formidable challenge for laser interferometers. In SOGRO, the vertical strain component h'_{33} largely makes up for this deficiency. With the aid of the local strain channels, the NN has been rejected to 10^{-3} for *all* θ.

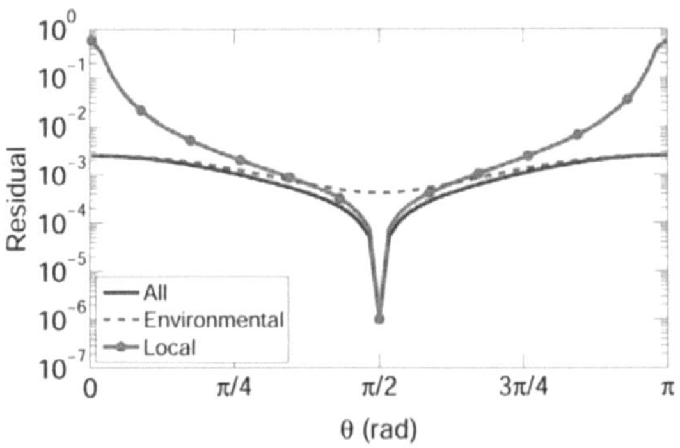

Figure 1. NN due to Rayleigh waves removed to ~10^{-3} by using h'_{13}, h'_{23}, h'_{33} and a_z (vertical CM), plus seven seismometers with SNR = 10^3 at the radius of 5 km.

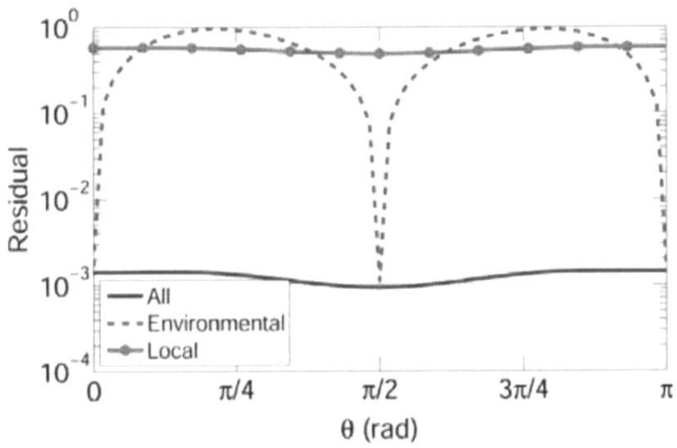

Figure 2. NN due to infrasound removed to 10^{-3} by using h'_{13}, h'_{23}, h'_{33} and 15 microphones of SNR = 10^4, one at the detector, seven each at the radius of 600 m and 1 km.

3. Mitigation of NN on interferometers with the aid of SOGROs

Since SOGRO is a very sensitive gravity strain gauge, one may be able to employ scaled-down SOGROs with arm-length $\ell \ll L$, in place of a large array of seismometers, to directly measure and remove the NN affecting the interferometer test masses. We restrict our discussion to underground detectors like KAGRA [3] or Einstein Telescope (ET) [4].

The Rayleigh waves are expected to dominate the NN for an underground detector [1]. In the presence of a GW and Rayleigh waves, the arm-length along the x axis is modulated by

$$\Delta L = hL + X_R(x_2) - X_R(x_1), \tag{6}$$

where $X_R(x_i)$ is the first term of equation (1) summed over multiple waves for the i-th test mass on the x axis. At 10 Hz, the Rayleigh wave length becomes λ_R ~ 350 m << L, causing $X_R(x_1)$ and $X_R(x_2)$ to be uncorrelated. Hence we need to measure $X_R(x_i)$ for each test mass by using a separate SOGRO co-located with it, as shown in figure 3.

From equations (1) and (2), we find that the 13-component of the SOGRO output is given by

$$\eta_{13}(x_i) = h_{13} + (i\omega / c_R) X_R(x_i). \tag{7}$$

We solve equation (7) for $X_R(x_i)$ and substitute it into equation (6) to obtain

$$h = \frac{\Delta L}{L} - i\frac{c_R}{\omega L}\left[\eta_{13}(x_2) - \eta_{13}(x_1)\right]. \tag{8}$$

The sensitivity required for mini-SOGRO to recover h is then given by

$$\eta = \frac{\Delta \ell}{\ell} = \frac{1}{\sqrt{2}}\frac{\omega L}{c_R} h. \tag{9}$$

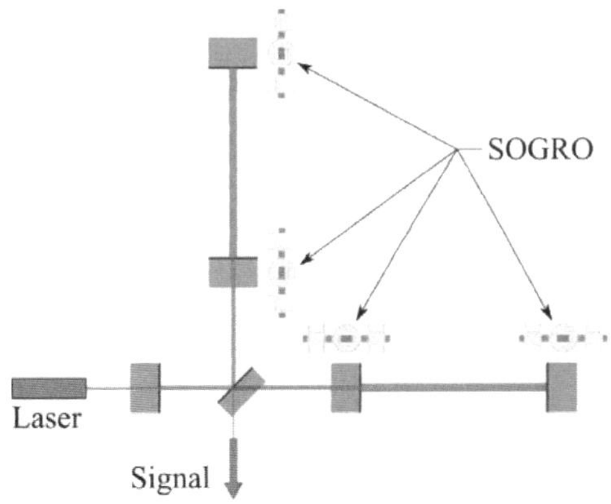

Figure 3. Four mini-SOGROs collocated with four test masses of a laser interferometer GW detector.

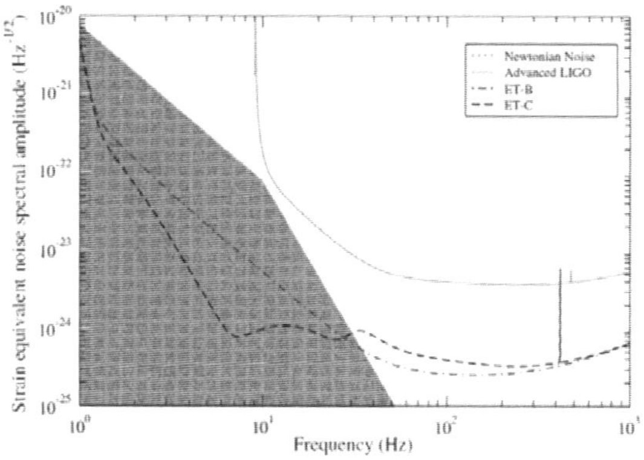

Figure 4. Sensitivity goals of aLIGO and ET. The shaded region represents the parameter space dominated by the NN.

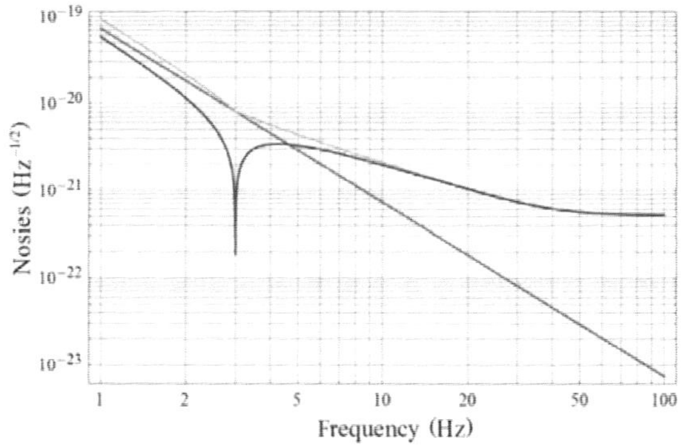

Figure 5. Instrument noise spectral density of mini-SOGRO of 5-m arm-length cooled to 0.1 K and coupled to a nearly quantum-limited SQUID amplifier.

Figure 4 shows the sensitivity goals of advanced LIGO (aLIGO) [5] and ET [4]. The shaded region represents the parameter space dominated by the NN at a shallow depth. A worthy goal would be rejecting the NN by an order of magnitude to $h \approx 10^{-22}$ Hz$^{-1/2}$ at 3 Hz and to 10^{-23} Hz$^{-1/2}$ at 10 Hz. For ET with L = 10 km, equation (9) yields $\eta \approx 4 \times 10^{-21}$ Hz$^{-1/2}$ at 3 Hz and 1.3×10^{-21} Hz$^{-1/2}$ at 10 Hz. The NN between SOGRO test masses must be highly correlated. According to Beker *et al.* [6],

$$S = \frac{1}{\sqrt{1-C_{SN}^2}} \le \frac{c_R}{\omega \ell}, \tag{10}$$

where S is the mitigation factor and C_{SN} is the correlation between the test masses. To obtain S = 10 at 10 Hz, we need C_{SN} = 0.995 and $\ell \le c_R/\omega S$ = 5.6 m. Mitigating the NN for ground detectors is much more challenging since the low speed of the Rayleigh waves on the surface, $c_R \approx 250$ m/s, reduces ℓ to ≤ 0.4 m. Such a small SOGRO would hardly have enough sensitivity.

Figure 5 shows the instrument noise spectral density for SOGRO with ℓ = 5 m, M = 1 ton (each test mass), and $Q = 5 \times 10^8$ cooled to 0.1 K and coupled to a dc SQUID with $2\hbar$ noise. The expected sensitivity of SOGRO comes to within a factor of 2 from that required for S = 10. The same SOGRO with $Q = 10^9$ coupled to a $1\hbar$ SQUID would meet the sensitivity requirement, provided all the other noise could be reduced to below its intrinsic noise limit. Should these sensitivities be achieved, SOGRO could make it possible to construct ET less deep.

It is interesting to see how a mini-SOGRO two orders of magnitude less sensitive to GWs can help ET mitigate the NN by an order of magnitude. This is because a SOGRO with ℓ = 5 m is quite efficient to detect the Rayleigh NN with $\lambda_R/2\pi$ = 56 m and SOGRO employs a highly sensitive superconducting displacement sensor. Although achieving a test mass Q of 10^9 and reaching the quantum limit for the SQUID noise is very challenging, it is worth investigating the SOGRO option since it has intrinsic advantages over seismometers in that it detects the NN directly and can monitor the local gravity gradient environment with high sensitivity.

[1] Harms J *et al* 2013 *Phys.* Rev. *D* **88** 122003
[2] Harms J and Paik H J 2015 *Phys. Rev. D* **92** 022001
[3] Somiya K 2012 *Class. Quantum Grav.* **29** 124007
[4] ET Science Team 2011 *Einstein Gravitational Wave Telescope Conceptual Design Study*, available from European Gravitational Observatory http://www.et-gw.eu/. ET-0106C-10.
[5] LIGO Scientific Collaboration 2009 *Rep. Prog. Phys.* **72** 076901
[6] Beker M G *et al* 2011 *Gen. Rel. Grav.* **43** 623

Journal of Physics: Conference Series **716** (2016) 012026 doi:10.1088/1742-6596/716/1/012026

Determination of mass of an isolated neutron star using continuous gravitational waves with two frequency modes: an effect of a misalignment angle

Kazunari Eda[1,2], Kenji Ono[1,3], and Yousuke Itoh[2]

[1] Department of Physics, Graduate School of Science, University of Tokyo, Tokyo, 113-0033, Japan

[2] Research center for the early universe, Graduate School of Science, University of Tokyo, Tokyo, 113-0033, Japan

[3] Institute for cosmic ray research, University of Tokyo, 5-1-5 Kashiwanoha, Kashiwa, Chiba 277-8582, Japan

E-mail: eda@resceu.s.u-tokyo.ac.jp

Abstract. A rapidly spinning neutron star (NS) would emit a continuous gravitational wave (GW) detectable by the advanced LIGO, advanced Virgo, KAGRA and proposed third generation detectors such as the Einstein Telescope (ET). Such a GW does not propagate freely, but is affected by the Coulomb-type gravitational field of the NS itself. This effect appears as a phase shift in the GW depending on the NS mass. We have shown that mass of an isolated NS can, in principle, be determined if we could detect the continuous GW with two or more frequency modes. Indeed, our Monte Carlo simulations have demonstrated that mass of a NS with its ellipticity 10^{-6} at 1 kpc is typically measurable with precision of 20% using the ET, if the NS is precessing or has a pinned superfluid core and emits GWs with once and twice the spin frequencies. After briefly explaining our idea and results, this paper concerns with the effect of misalignment angle ("wobble angle" in the case of a precessing NS) on the mass measurement precision.

1. Introduction

The distribution of masses of neutron stars (NSs) gives insights to understand their birth mechanisms and evolution histories. For instance, the mass measurements of the massive pulsar PSR J16142230 [1] and PSR J0348+0432 [2] have great impacts on exotic matter physics and studies of NS interiors.

Currently, mass measurements of NSs are mostly limited to those in binaries [3]. Since possible mutual mass transfers in binaries may change the mass distributions of the component NSs, it is desirable to measure masses of as many isolated NSs as possible.

We have proposed a new method to measure mass of an isolated NS by detecting Coulomb-type phase shifts imprinted in gravitational waves (GWs) by the NS gravitational field [4]. In that paper, we have shown that the mass of a NS at 1 kpc distance and with its ellipticity 10^{-6} is typically measurable with precision of 20% using the third generation GW telescope such as the Einstein Telescope (ET) [5], if the NS is precessing or has a pinned superfluid core and emits GWs with once and twice the spin frequencies.

In the following sections, we will review the idea of our method, and then comment on our assumptions on misalignment angles in our original paper [4], namely the misalignment angle of the freely-precessing NSs and their effect on mass measurement precision.

2. Method

Let us consider an isolated NS with mass M located at the coordinate origin $r = 0$. We assume that the NS emits GW with multiple frequencies $\omega_\alpha = 2\pi f_\alpha (\alpha = 1, 2, \cdots)$. The two polarization modes of GWs from the NS can be written as

$$\bar{h}_+ = \sum_\alpha A_{+,\alpha} \cos \Psi_\alpha, \quad \bar{h}_\times = \sum_\alpha A_{\times,\alpha} \sin \Psi_\alpha, \tag{1}$$

$$\Psi_\alpha \equiv -\omega_\alpha t + \omega_\alpha r + \Phi_\alpha, \tag{2}$$

$$\Phi_\alpha \equiv 2\omega_\alpha M \ln[2\omega_\alpha r] + \phi_{R\alpha}, \tag{3}$$

where $\phi_{R\alpha}$ denotes the constant reference GW phases. The logarithmic term in Φ_α is the Coulomb phase shift due to the static part of the NS gravitational field [4, 6]. All the searches for GWs from rapidly rotating NSs conducted so far have not taken into account these Coulomb phase shifts.

Our method to measure the mass M of a NS and results are as follows [4]. Suppose that we determine GW phases Φ_α for two or more frequency modes (say, ω_1 and ω_2 with the ratio $K \equiv \omega_2/\omega_1$). If these two modes have such a property that

$$\phi_{R2} - K\phi_{R1} = 0, \tag{4}$$

we can determine the mass from the following combination,

$$\Phi_2 - K\Phi_1 = 2\omega_2 M \ln K \simeq 0.06 \left(\frac{M}{1.4M_\odot}\right)\left(\frac{f_2}{1\mathrm{kHz}}\right)\left(\frac{\ln K}{\ln 2}\right). \tag{5}$$

The condition (4) is satisfied for a freely-precessing isolated bi-axial NS, which may emit GWs (approximately) at once and twice the spin frequencies ($K = 2$). Another possibility can be found in the model proposed by Jones [7]. He showed that if a tri-axial NS contains a pinned superfluid core and none of the axes of the principal moments of inertia of the solid crust aligns with the spin angular velocity vector of the core, then such a NS emits GWs at once and twice the spin frequencies. Notably, the condition (4) is hold when such a NS is bi-axial (but none of the axes of the principal moments of inertia is aligned with the spin axis).

The GWs from these two models of NSs have (See [4, 8, 9, 10, 11])

$$A_{+,1} = \frac{1}{4} h_0 \sin 2\theta \sin \iota \cos \iota, \qquad A_{\times,1} = \frac{1}{4} h_0 \sin 2\theta \sin \iota, \tag{6}$$

$$A_{+,2} = \frac{1}{2} h_0 \sin^2 \theta \left(1 + \cos^2 \iota\right), \quad A_{\times,2} = h_0 \sin^2 \theta \cos \iota, \tag{7}$$

where the inclination angle ι is defined as the angle between the rotational axis and the line-of-sight, and the misalignment angle θ is defined as the angle between the principal axis and the angular momentum axis. The overall amplitude is $h_0 = 4\varepsilon I \omega_1^2/r$ where I denotes the star's average moment of inertia and the ellipticity ε quantifies the degree of the non-axisymmetry.

Assuming these models, we have performed Monte-Carlo simulations to study the measurement precision of NS mass using Eq. (5) for three-year ET observation [4]. For those simulations, we have randomly selected sets of waveform parameters $(\theta, \iota, \phi_R, \psi, \alpha, \delta)$ where ϕ_R is the reference phase, ψ is the GW polarization phase, and α and δ are the sky position of the NS. Uniform distributions are assumed for those parameters. The NS moment of inertia I and its ellipticity ε are set to be 10^{45} g$\cdot$cm^2 and 10^{-6} [12, 13], respectively. We adopted several representative values of r (1 kpc, 10 kpc, and 50 kpc) and f_1 (300 Hz and 500 Hz). Then, We estimate the measurement error in the NS mass for each simulation.

In the ideal case where all the waveform parameters except for ϕ_R and M are known, the measurement error in the GW phase Φ_α is $1/\rho$ where ρ is the signal-to-noise ratio (SNR). Hence, GW detections with $\rho \gtrsim 100$ for both modes may suffice to estimate the NS mass as indicated in Eq. (5). When all the waveform parameters are unknown in advance, the correlations among the parameters degrade the mass measurement precision. To improve the mass measurement precision, we have assumed that the spin frequency, the spin-down rate, the sky position of the NS are known in advance from electromagnetic observations or GW observations by the second generation GW detectors such as the advanced LIGO [14], advanced Virgo [15], and KAGRA [16].

The resulting cumulative distributions of mass measurement precision are plotted in the Fig. 1 of [4]. For example, we found that the mass of the NS with its spin frequency 500 Hz and its ellipticity is typically measurable with an accuracy of 20% using the ET.

3. Misalignment angle

We have assumed a uniform distribution for the misalignment angle θ in our Monte Carlo simulations in the previous paper [4]. In the case of a NS containing a pinned superfluid core, there seems to be neither observational nor theoretical constraint on θ. In fact, a uniform distribution for the misalignment angle θ is assumed in this model [9, 10, 11].

A possible range of the misalignment angle θ in the case of a freely precessing NS is not well-known. Two known examples indicate smaller misalignment angles: $\theta \simeq 3$ degrees (PSR B1828-11) [17] and $\theta \simeq 0.8$ degrees (PSR B1642-03) [18]. On the other hand, the theoretical maximum possible value of θ is estimated by Eq. (24) of [19] as $\theta_{\rm max} \simeq 10$ degrees $(500{\rm Hz}/f_1)(u_{\rm break}/10^{-2})$ where $u_{\rm break}$ is the breaking strain of the NS crust. For the freely precessing magnetars recently discovered in [20, 21], parameter degeneracies prevent from inferring misalignment angles. In any case, one may find it questionable to use a uniform distributions for θ in the case of a precessing NS.

To see how the misalignment angle affects the measurement precision of NS mass in our method, we have conducted 10,000 Monte Carlo simulations for three-year ET observations similar to the previous paper, but for several fixed misalignment angles. In these simulations, the sky position, polarization angle, reference phase, and inclination angles are randomly chosen from uniform distributions. The distance of $r = 1$ kpc and spin frequency of $f = 500$ Hz are assumed. Fisher matrix method is used to estimate measurement errors of the waveform parameters. As in the previous simulations, the frequency, spin-down rate, sky position of each NS are assumed to be known. The results are shown in Fig. 1. This figure indicates that masses of more than 70% of NSs are measurable with a precision of $\Delta M/M \simeq 0.2$ if $10^\circ \leq \theta \leq 80^\circ$. On the other hand, if the misalignment angles are smaller than 10 degrees, as suggested by PSR B1828-11 and PSR B1642-03, even the third generation GW telescope cannot determine the NS mass with sufficient precision (e.g. $\Delta M/M \simeq 0.8$ for 20% of NSs in the case of $\theta = 3^\circ$). This is because the amplitudes for both modes become smaller for the smaller misalignment angle as indicated by Eqs. (6) and (7).

4. Acknowledgments

We thank Toshio Nakano for informing us that the misalignment angles are not determined for the freely precessing magnetars [20, 21] due to parameter degeneracy. We also thank Bruce Allen for pointing out that mass measurement precision would be degraded for freely precessing stars with θ close to zero. This work is supported by JSPS Fellows Grant No. 26.8636 (K. E.),

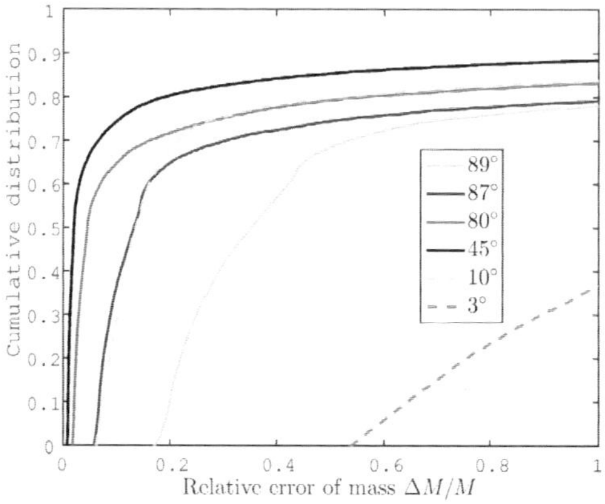

Figure 1. The cumulative distribution functions for relative measurement errors in NS mass $\Delta M_{\rm NS}/M_{\rm NS}$. This figure is obtained by Monte Carlo simulations where the NS sky position, polarization angle, reference GW phase, and inclination angles are randomly chosen from uniform distributions. The distance of $r = 1$ kpc, spin frequency of $f = 500$Hz, and three year observation by the ET are assumed. The solid lines correspond to NSs with the misalignment angles of $\theta = 45°$ (black line), 80° (red line), 87° (blue line), 89° (green line), while the dashed lines correspond to $\theta = 10°$ (cyan line) and 3° (magenta line), respectively. While a NS with a pinned superfluid core may have any values of θ, a freely precessing NS may be limited by $\theta < 10$ degrees. Mass measurement precision for a freely precessing NS may be worse than that for a NS with a pinned superfluid core.

JSPS Grant-in-Aid for Young Scientists Grant No. 25800126, and the MEXT Grant-in-Aid for Scientific Research on Innovative Areas (Grant Number 24103005) (Y. I.).

References

[1] Demorest P B *et al.* 2010 *Nature* **467** 1081–1083
[2] Antoniadis J *et al.* 2013 *Science* **340** 448
[3] Lattimer J M 2011 *Astrophys. Space Sci.* **336** 67–74
[4] Ono K, Eda K and Itoh Y 2015 *Physical Review D* **91** 084032
[5] Punturo M *et al.* 2010 *Classical and Quantum Gravity* **27** 194002
[6] Asada H and Futamase T 1997 *Physical Review D* **56** 6062–6066
[7] Jones D I 2010 *Mon. Not. R. Astron. Soc.* **402** 2503–2519
[8] Jaranowski P, Królak A and Schutz B F 1998 *Physical Review D* **58** 063001
[9] Bejger M and Królak A 2014 *Classical and Quantum Gravity* **31** 105011
[10] Jones D I 2015 *Mon. Not. R. Astron. Soc.* **453** 53–66 (*Preprint* 1501.05832)
[11] Pitkin M *et al.* 2015 *Mon. Not. R. Astron. Soc.* **453** 4399–4420
[12] Horowitz C J and Kadau K 2009 *Physical Review Letters* **102** 191102
[13] Hoffman K and Heyl J 2012 *Mon. Not. R. Astron. Soc.* **426** 2404–2412
[14] Abbott B *et al.* (LIGO Scientific Collaboration) 2009 *Rept.Prog.Phys.* **72** 076901
[15] Accadia T *et al.* (VIRGO Collaboration) 2012 *Journal of Instrumentation* **7** P03012
[16] Aso Y *et al.* (KAGRA Collaboration) 2013 *Physical Review D* **88** 043007
[17] Link B and Epstein R I 2001 *Astrophysical Journal* **556** 392–398
[18] Shabanova T V, Lyne A G and Urama J O 2001 *Astrophysical Journal* **552** 321–325
[19] Jones D I and Andersson N 2002 *Mon. Not. R. Astron. Soc.* **331** 203–220
[20] Makishima, K *et al.* 2014 *Physical Review Letters* **112** 171102
[21] Makishima K *et al.* 2015 *PASJ Advance Access* psv097

TorPeDO: A Low Frequency Gravitational Force Sensor

D. J. McManus,[1] M. J. Yap,[1] R. L. Ward,[1] D. A. Shaddock,[1] D. E. McClelland,[1] B. J. J. Slagmolen[1]

[1]Department of Quantum Science, Australian National University,Building 38a, Science Road, Acton, ACT 0200, Australia

E-mail: `david.mcmanus@anu.edu.au`

Abstract. Second generation gravitational wave detectors are likely to be limited by Newtonian Noise at low frequencies. A dual torsion pendulum sensor aimed at exploring low-frequency gravitational-force noise is being studied at the ANU. This sensor is designed to measure local gravitational forces to high precision and will be limited by Newtonian noise. We report on a controls prototype which has been constructed and suspended, along with initial characterisation and testing of the two torsion pendulums. Large weights at the end of each bar reposition the centres of mass to the same point in space external to both bars. Since both bars have a common suspension point, resonant frequency (≈ 33.4 mHz) , and centre of mass, mechanical disturbances and other noise will affect both bars in the same manner, providing a large mechanical common mode rejection.

1. Introduction

Second generation gravitational wave detectors are about to begin taking measurements, with strain sensitivities as low as $3 \times 10^{-24}/\sqrt{\text{Hz}}$[1] in the frequency range from 10 Hz to 10 kHz. At lower frequencies these detectors are potentially limited by seismic and atmospherically induced Newtonian Noise[2]. To understand and characterise Newtonian Noise, a direct measurement of Newtonian Noise is first required.

The Torsion Pendulum Dual Oscillator (TorPeDO) is a low frequency gravitational force sensor which aims to be limited by Newtonian Noise. The system features two torsion pendulums which are suspended along the same axis of rotation and perpendicular to each other, similar to the TOBA design[3]. Local gravity gradient changes cause a differential attraction on the two torsion bars which causes them to rotate in opposite directions. A differential optical measurement is taken between the two torsion bars to measure this. The mechanical configuration is chosen to maximise common mode rejection between the torsion bars. This is achieved by locating the centres of mass of both bars to the same point in space and tuning the torsion bars to the same frequency for most modes.

Other low frequency gravitational forces can be measured by this system, making it useful for early earthquake warning, giving advanced warning over seismometers for an earthquake 50 km away[4]. Low frequency systems such as the TorPeDO are of interest for testing theories of semi-classical gravity[5], by looking for a semi-classical resonance offset from the main resonance. A

Journal of Physics: Conference Series **716** (2016) 012027 doi:10.1088/1742-6596/716/1/012027

smaller version of this sensor is also planned for construction to measure quantum radiation pressure noise.

A controls prototype for this sensor has been constructed and suspended. This prototype is currently in air with a static suspension as a proof of concept; to demonstrate our control and measurement method. It will later be placed in vacuum with suspension improvements to lower noise contributions and increase force sensitivity.

2. Prototype Design

The controls prototype for the TorPeDO features two 60 cm long aluminium bars. Large steel weights are positioned at the ends of each bar to reposition the centres of mass to the same point in space, external to both bars. Each bar is suspended by two straight tungsten wires with a separation of 23 mm. This double wire suspension allows for both bars to easily be suspended with the same axis of rotation. While a double wire suspension raises the frequency of the torsional mode, we still achieve a suitably low resonance frequency of roughly 33 mHz for our design. Cavity mirrors are mounted at the end of the bars in line with the centre of mass at a 45° angle. Once the two bars are suspended, these mirrors form 4 linear optical cavities around the perimeter of the system as shown in Figure 1. Figure 2 shows a picture of the constructed prototype at ANU.

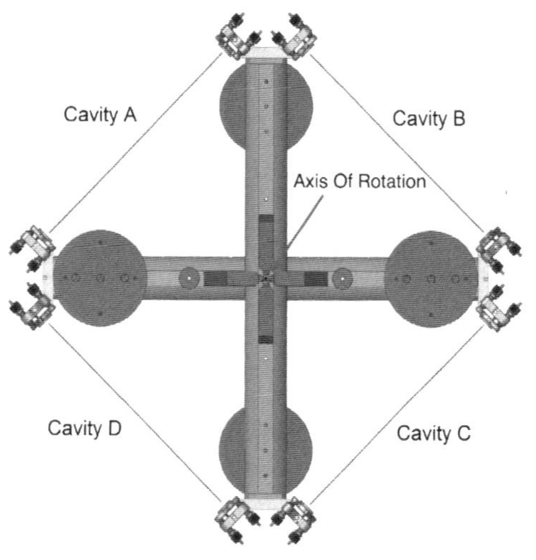

Figure 1. Top down engineering drawing of the controls prototype design

Figure 2. Photo of the controls prototype suspended at ANU.

The experiment is currently in air, and is limited by environmental disturbances. The prototype will be moved inside vacuum along with an isolated suspension platform to improve sensitivity. A planned upgrade to the prototype will replace the tungsten suspension wires with fused silica, reducing thermal noise levels down to an estimated $10^{-18}/\sqrt{\text{Hz}}$ strain at 1 Hz. After this upgrade, the estimated sensitivity level of the TorPeDO prototype is $10^{-16}/\sqrt{\text{Hz}}$ strain at 1 Hz. The 60 cm bar length TorPeDO prototype is predicted to be sensitive to Newtonian noise below 0.1 Hz.

3. Tuning Masses

Tuning masses are used in the mechanical design to allow for small alterations to be made to physical mass properties of the system. The first set of tuning masses are positioned on top of both bars on either side of the axis of rotation. These masses can be moved up and down along a thread to shift the location of the centre of mass for that bar. This also allows tuning of the pendulum and pitch modes of the bars. A second set of tuning masses are positioned in a hollow channel that runs through the centre of the bar. These masses can be moved inwards and outwards, which changes the moment of inertia of the bars about the torsional axis. This allows for tuning of the torsional frequency of the bars. Correct positioning of these masses allows for the two frequencies to be matched as closely as possible. The frequency matching goal for this prototype is a match to within 1 part in 1000 to achieve the sensitivity goal quoted in the previous section. By mismatching the distance of these tuning masses from the axis of rotation, the centre of mass can also be tuned along the bar axis. The centres of mass must be well tuned to the rotation axis on each bar to minimise coupling from suspension point displacement into yaw rotation. This offset should be reduced to less than 1 mm to prevent this coupling from limiting our sensitivity.

4. Sensing and Actuation

In the current controls prototype, two different systems are used for sensing and actuation. A pair of optical levers and shadow sensors are used to measure rotation of the torsion bars. BOSEM coils are used to actuate and provide feedback to the bars. These systems allow for characterisation of individual bar motion and to control the differential torsional mode.

5. Transfer Functions

Transfer functions have been taken of both bars by driving through the BOSEM actuators with white noise and measuring the response with the optical levers and shadow sensors. Figure 3 shows the transfer functions of both bars measured using the shadow sensors. The low frequency peak on both plots corresponds to the torsional resonance of the bars (≈ 33 mHz). The second main feature at ≈ 0.6 Hz corresponds to the pendulum mode of the bars.

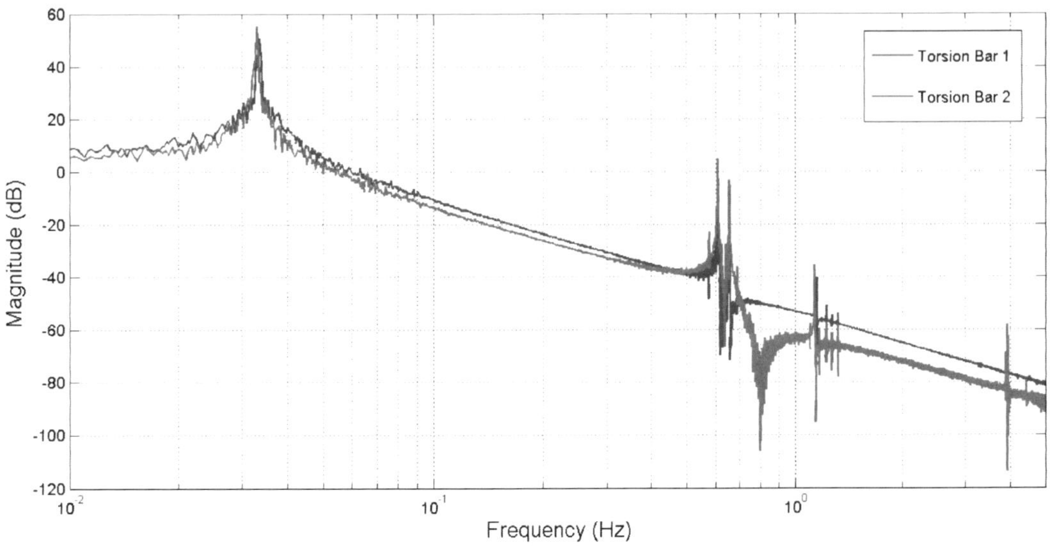

Figure 3. White noise transfer functions.

6. Conclusion and Future Work

A controls prototype of the TorPeDO has been constructed and suspended at ANU. The next stage of this experiment involves further characterisation of the mechanical properties of the prototype, including the tuning range and quality factor of the torsional and pendulum modes. The optical read-out using the arm cavities of the TorPeDO must also be implemented before the system is placed in vacuum.

References

[1] G. M. Harry et al., CQG 27, 084006 (2010).
[2] J. Harms, et al. Phys. Rev. D 88, 122003 (2013)
[3] M. Ando, et al. Phys. Rev. Lett., 105(16), 2010
[4] J. Harms, et al. Geophys J Int, 201(3) (2015).
[5] H. Yang, et. al., Phys. Rev. Lett., 110, 170401 (2013)

Augmented kludge waveforms and Gaussian process regression for EMRI data analysis

Alvin J K Chua

Institute of Astronomy, University of Cambridge, Madingley Road, Cambridge CB3 0HA, United Kingdom

E-mail: ajkc3@ast.cam.ac.uk

Abstract. Extreme-mass-ratio inspirals (EMRIs) will be an important type of astrophysical source for future space-based gravitational-wave detectors. There is a trade-off between accuracy and computational speed for the EMRI waveform templates required in the analysis of data from these detectors. We discuss how the systematic error incurred by using faster templates may be reduced with improved models such as augmented kludge waveforms, and marginalised over with statistical techniques such as Gaussian process regression.

1. Introduction

The end-2015 launch of the LISA Pathfinder satellite [1] is set to usher in a new era of space-based gravitational-wave (GW) detectors such as eLISA [2] and DECIGO [3]. Data collected from these detectors will contain a wealth of astrophysical information, which may be used to complement electromagnetic observations [4] and to test strong-field general relativity [5].

One important type of source for space-based detectors is the extreme-mass-ratio inspiral (EMRI) of a stellar-mass compact object (CO) into a massive black hole (BH). The inspiral exhibits strong relativistic effects a few years before the CO plunges; these are imprinted on the GW signal from the source. Matching the signal to a waveform template from an accurate EMRI model will then allow precise estimation of the source's astrophysical parameters.

Accurate EMRI waveforms based on the Teukolsky equation [6] and gravitational self-force calculations [7] are available, but are extremely expensive to generate. The analysis of detector data will require computationally affordable alternatives such as the analytic kludge (AK) [8] and numerical kludge (NK) [9] models, which combine mixed prescriptions for computing the orbit and waveform. However, the systematic error in these templates with respect to the true GW signal will lead to reduced detection prospects and inaccurate parameter estimation.

These proceedings are based on a talk given by the author at the 11th Edoardo Amaldi Conference on Gravitational Waves. In this talk, we discussed methods for reducing and marginalising over model error in EMRI waveforms. Sec. 2 introduces an augmented kludge waveform that significantly improves on the accuracy of the AK model, with virtually no added computational cost [10]. It uses a frequency map to the three fundamental frequencies associated with a geodesic orbit around a Kerr black hole [11], and features updated evolution equations along with a fit to the slower but more accurate NK model.

While the trade-off between accuracy and computational speed for EMRI waveforms may be lessened with improved models, it cannot be removed completely. However, the statistical technique of Gaussian process regression (GPR) may be used to marginalise over the systematic error in a GW model [12], which significantly improves parameter estimation performed with faster but less accurate templates [13]. The application of GPR to EMRI data analysis is discussed in Sec. 3, where its feasibility is illustrated with a one-parameter likelihood search.

2. Augmented kludge waveforms

Kludge waveform models are semi-relativistic EMRI models designed for robust use in data analysis. The inspiral trajectory in the NK model of Babak et al. [9] is built from Kerr geodesics and evolved with post-Newtonian (PN) expressions. On orbital timescales, the motion of the CO is approximately geodesic and hence features periapsis precession of the orbital ellipse, along with Lense–Thirring precession of the orbital plane. Both forms of precession are due to discrepancies in the frequencies $\omega_{r,\theta,\phi}$ of the radial, polar and azimuthal components of motion; expressions for these frequencies in terms of the (quasi-Keplerian) orbital parameters (e, ι, p) of the geodesic have been given by Schmidt [11].

The inspiral trajectory in the AK model of Barack and Cutler [8] is quicker to compute but less accurate, as it consists instead of PN-evolved Keplerian ellipses parametrised by (e, ι, p). The orbital frequency f_{orb} is determined by Kepler's third law, while the two precession rates f_{peri} and f_{LT} are introduced and evolved separately with additional PN expressions. However, $(f_{\mathrm{orb}}, f_{\mathrm{peri}}, f_{\mathrm{LT}})$ do not in general equal the correct values $(\omega_r, \omega_\phi - \omega_r, \omega_\phi - \omega_\theta)$. This causes the AK waveform to dephase rapidly relative to the NK waveform, even at the early-inspiral stage.

We propose an augmented AK model [10] that matches the three AK frequencies with the appropriate combinations of $\omega_{r,\theta,\phi}$; this is achieved by mapping the BH mass M, the BH spin parameter a and the semi-latus rectum p to some unphysical values $(\tilde{M}, \tilde{a}, \tilde{p})$, which are given implicitly by the algebraic system of equations

$$(f_{\mathrm{orb}}, f_{\mathrm{peri}}, f_{\mathrm{LT}})|_{(\tilde{M},\tilde{a},\tilde{p})} = (\omega_r, \omega_\phi - \omega_r, \omega_\phi - \omega_\theta)|_{(M,a,p)}. \tag{1}$$

Substituting $(\tilde{M}, \tilde{a}, \tilde{p})$ for the physical parameters (M, a, p) in the AK model provides an instantaneous correction of the three AK frequencies at any point along the inspiral trajectory.

In the augmented AK model, we evaluate the map at a single trajectory point (e_0, ι_0, p_0) to obtain the corresponding point $(e_0, \iota_0, \tilde{p}_0)$ on an unphysical trajectory, then evolve the unphysical trajectory as well as the three AK frequencies with higher-order 3PN $\mathcal{O}(e^6)$ expressions given by Sago and Fujita [14]. We also compute two additional points next to (e_0, ι_0, p_0) on the NK trajectory and evaluate the map at each point, which allows us to perform a local quadratic fit of the unphysical trajectory to the (mapped) NK trajectory for greater accuracy. As a result of this fit, the augmented model features quadratic-in-time evolution of the unphysical BH mass and spin parameters. Since the map-and-fit is only performed locally on the inspiral trajectory, the added computational cost is fixed but essentially insignificant.

For a generic early-inspiral EMRI with a CO mass of $\mu = 10\,M_\odot$, a BH mass of $M = 10^6\,M_\odot$, a BH spin of $a = 0.8M$, an initial eccentricity of $e_0 = 0.5$, an initial inclination of $\iota_0 = \pi/6$ and an initial semi-latus rectum of $p_0 = 15M$, the original AK waveform and the NK waveform are a full cycle out of phase within three hours (see top plot of Fig. 1). The severe dephasing on this timescale is removed by the instantaneous correction of the three AK frequencies in the augmented model (see bottom plot of Fig. 1). For late-inspiral EMRIs over longer timescales, the improvement in accuracy provided by the augmented AK waveform is just as pronounced. We consider several sources with differing μ, a and e_0, choosing p_0 for each EMRI such that the CO plunges at $t = 1\,\mathrm{yr}$ [10]; the length of time over which the AK waveform remains phase-coherent with the NK waveform is increased from under an hour (in the original model) to over two months (in the augmented model) for all the sources.

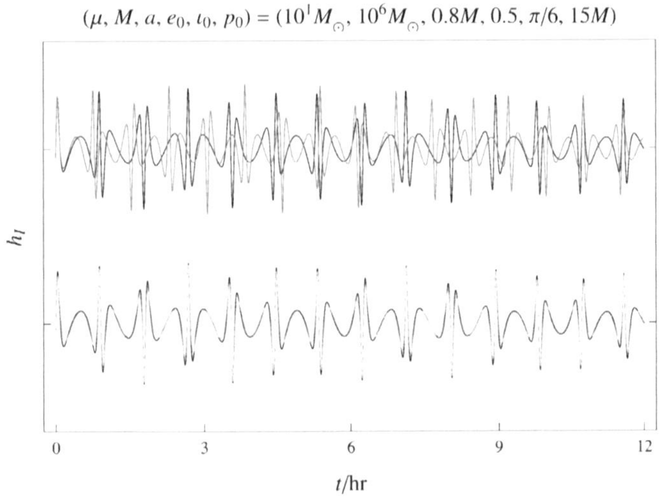

Figure 1. First 12 hours of original (red) and augmented (green) AK waveforms overlaid on NK waveform (black), for generic EMRI with initial semi-latus rectum of $15M$.

3. Gaussian process regression

In GW data analysis, waveform templates $H(\lambda)$ are used to construct the Bayesian likelihood $L(\lambda|x)$ for the model parameters λ, given the data x. However, any theoretical error in the waveform model H will reduce detection signal-to-noise ratios (SNRs) and parameter estimation accuracy. In a recently proposed method for dealing with model error [12], the likelihood is marginalised over the difference δh between the approximate waveform H and an accurate (true) waveform model h. This marginalisation is performed analytically by taking the prior for δh to be a Gaussian process trained on a small pre-computed set of differences $\{\delta h(\lambda_1), \ldots, \delta h(\lambda_n)\}$. The new GPR likelihood is still cheap to compute (since it depends only on H and the small training set), and will significantly improve parameter estimation prospects for Advanced LIGO and other ground-based GW detectors [13].

Interpolation of the waveform difference δh from the training set is central to the GPR method. Hence the performance of the marginalised likelihood depends strongly on the spacing between neighbouring training set points, with the maximal characteristic spacing $\Delta\lambda$ (beyond which the interpolant becomes suboptimal) determined by the accuracy of the approximate model H. For GPR to be useful in EMRI data analysis, $\Delta\lambda$ must be significantly larger than the characteristic grid size $\Delta\lambda_{\mathrm{TB}}$ in a template bank search with the accurate model h. In the worst case of an inaccurate approximate model where δh has comparable SNR to H and h, we have $\Delta\lambda \sim \Delta\lambda_{\mathrm{TB}}$ for an SNR of 1. However, $\Delta\lambda_{\mathrm{TB}}$ scales inversely with SNR while $\Delta\lambda$ does not, and so the number of training set points required to estimate the parameters of a typical eLISA source with an SNR of 30 will be $\sim 10^9$ times smaller than the number of template bank points required ($\sim 10^{30}$ for a fully coherent search in the 14-dimensional parameter space [15]).

To illustrate the feasibility of the GPR method for EMRI parameter estimation, we apply it to a one-dimensional search in the BH mass M, using the NK and original AK waveforms as accurate and approximate templates respectively. The true signal is taken to be an NK waveform with $M/M_\odot = M_T = 10^6$ and an SNR of 30. Fig. 2 shows the accurate (true) and approximate likelihoods, as well as two marginalised likelihoods with sparse ($\Delta M/M_\odot \approx 20$) and dense ($\Delta M/M_\odot \approx 10$) training sets. The true likelihood is sharply peaked at the correct value due to the large SNR, while there is no local peak in the approximate likelihood since the original AK model is too inaccurate. Both marginalised likelihoods show a peak at the

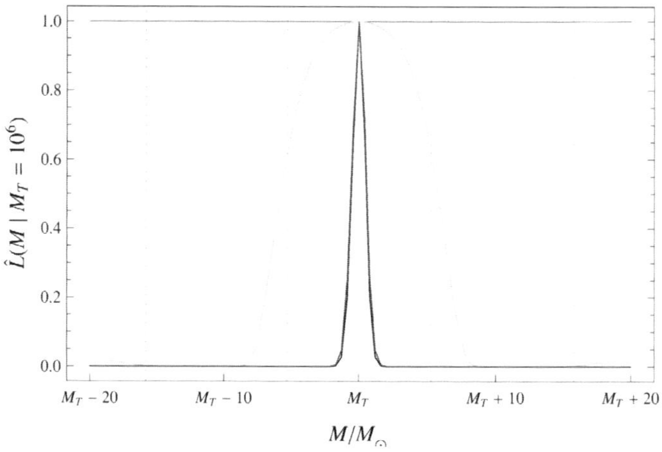

Figure 2. Approximate likelihood (red) and marginalised likelihoods with sparse (green) and dense (blue) training sets overlaid on true likelihood (black), for generic EMRI with BH mass of $10^6\,M_\odot$ and SNR of 30. Likelihoods are normalised to peak value of 1. Vertical lines mark positions of points in sparse (dashed) and dense (dotted) training sets.

correct value; the dense training set performs better and gives a near-perfect recovery of the true likelihood, which is far more expensive to sample. Although the two training sets considered in this example are still too dense for likelihood searches over the full parameter space, using improved approximate templates such as the augmented kludge waveforms in Sec. 2 should lower the training set density required for the marginalised likelihood to perform optimally.

4. Conclusion

We have proposed augmented kludge waveforms and Gaussian process regression as potential methods for reducing and marginalising over error in EMRI waveform models, with a view to improving detection and parameter estimation prospects for future space-based GW detectors.

Acknowledgements

This talk was based on work done in collaboration with Jonathan Gair and Christopher Moore at the Institute of Astronomy, Cambridge. The author's work was supported by the Cambridge Commonwealth, European and International Trust.

References

[1] Danzmann K 2015 *Nature Physics* **11** 613
[2] Amaro-Seoane P *et al.* 2012 *Classical and Quantum Gravity* **29** 124016
[3] Kawamura S *et al.* 2011 *Classical and Quantum Gravity* **28** 094011
[4] Sathyaprakash B S and Schutz B F 2009 *Living Reviews in Relativity* **12** 2
[5] Gair J R, Vallisneri M, Larson S L and Baker J G 2013 *Living Reviews in Relativity* **16** 7
[6] Hughes S A 2001 *Physical Review D* **64** 064004
[7] Barack L 2009 *Classical and Quantum Gravity* **26** 213001
[8] Barack L and Cutler C 2004 *Physical Review D* **69** 082005
[9] Babak S *et al.* 2007 *Physical Review D* **75** 024005
[10] Chua A J K and Gair J R 2015 *Classical and Quantum Gravity* **32** 232002
[11] Schmidt W 2002 *Classical and Quantum Gravity* **19** 2743
[12] Moore C J and Gair J R 2014 *Physical Review Letters* **113** 251101
[13] Moore C J, Berry C P L, Chua A J K and Gair J R 2016 *Physical Review D* (in press) arXiv:1509.04066[gr-qc]
[14] Sago N and Fujita R 2015 *Progress of Theoretical and Experimental Physics* **2015** 073E03
[15] Gair J R *et al.* 2004 *Classical and Quantum Gravity* **21** S1595

Construction of an optical test-bed for eLISA

Maike Lieser[1][2], E Fitzsimons[4][5], K-S Isleif[1][2], C Killow[3], M Perreur-Lloyd[3], D Robertson[3], S Schuster[1][2], M Tröbs[1][2], S Veith[1][2], H Ward[3], G Heinzel[1][2] and K Danzmann[1][2]

[1] Max Planck Institute for Gravitational Physics (Albert Einstein Institute), Callinstraße 38, 30167 Hannover, Germany
[2] Leibniz Universität Hannover, 30167 Hannover
[3] Scottish Universities Physics Alliance (SUPA), School of Physics and Astronomy, Institute for Gravitational Research, University of Glasgow, Glasgow, G12 8QQ, UK
[4] Airbus Defense and Space - Space Systems, 88039 Friedrichshafen, Germany
[5] present address: UK Astronomy Technology Centre, Royal Observatory Edinburgh, Edinburgh EH9 3HJ, UK

E-mail: maike.lieser@aei.mpg.de

Abstract. In the planned eLISA mission a key part of the system is the optical bench that holds the interferometers for reading out the inter-spacecraft distance and the test mass position. We report on ongoing technology development for the eLISA optical system like the back-link between the optical benches and the science interferometer where the local beam is interfered with the received beam from the distant spacecraft. The focus will be on a setup to investigate the tilt-to-pathlength coupling in the science interferometer. To test the science interferometer in the lab a second bench providing a laser beam and a reference interferometer is needed. We present a setup with two ultra-stable low expansion glass benches and bonded optics. To suppress the tilt-to-pathlength coupling to the required level (few μm/rad) imaging optics are placed in front of the interferometer photo diodes.

1. eLISA optical system technology development

'The Gravitational Universe' [1] was selected as the science theme for the L3 mission by the European Space Agency (ESA). The launch is currently planned for 2034. The Laser Interferometer Space Antenna (LISA), a laser interferometer measuring gravitational waves, addresses this science theme. At the Albert Einstein Institute in Hannover we are developing technologies for the optical system for eLISA (evolving LISA, standing for the constantly evolving LISA mission scheme). The experiments are addressing different aspects of the optical system that have been identified by the LISA community to need further experimental investigation.

2. eLISA interferometry test-bed

Together with the University of Glasgow and Airbus Defense and Space we developed and built a representative setup to test the interferometry for eLISA. The minimal optical bench operates with a reduced set of optical components to what the optical bench for an eLISA spacecraft would require. The optical bench is located between the test mass and the telescope for sending and receiving a laser beam from the distant spacecraft. It incorporates several interferometers for reading out the distance to the remote spacecraft and the test mass. The individual measurements are then combined to form the laser link between two test masses in distant spacecrafts.

The minimal version has only one measurement interferometer but is built with space qualified techniques and the experience from the University of Glasgow who built the optical bench for LISA Pathfinder [2]. The interferometer is the science interferometer that is reading out the inter spacecraft distance and this is the most challenging one because of spacecraft jitter and the long distance of millions of kilometers. The spacecraft jitter is inducing a tilt-to-pathlength coupling that is, after shot noise, the second largest noise source for eLISA. For reducing this coupling imaging optics are placed in front of the interferometer photo diodes. They are imaging the entrance pupil of the telescope to the photo diode to reduce the beam walk due to the tilt of the beam. There have been extensive theoretical studies on this topic [3] but no representative experiments have been conducted so far. Because of its significance a good understanding of the tilt-to-pathlength coupling and a verification of the simulations are important.

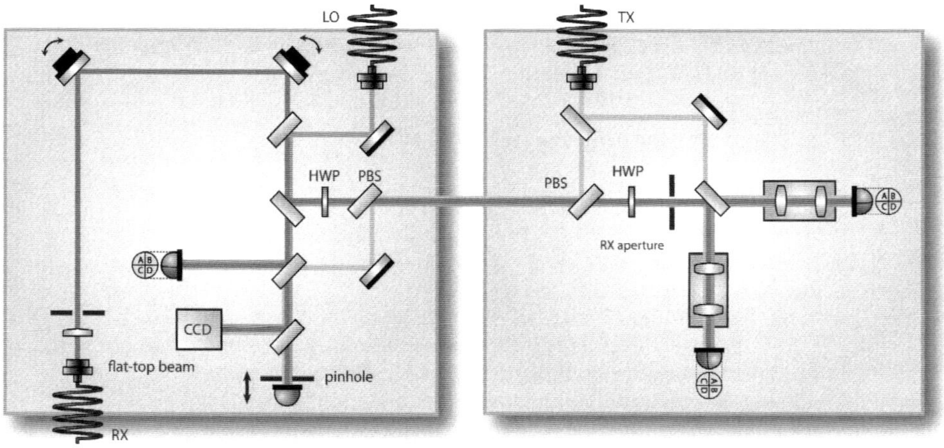

Figure 1. Schematic of the layout of the minimal Optical Bench (right) and the Telescope Simulator (left).

For operating the science interferometer a second bench, the so called Telescope Simulator, was built. The Telescope Simulator is providing a laser beam representing the beam from the distant spacecraft (RX beam). This beam is tilted for simulating spacecraft jitter and has a flat top beam profile. A schematic of the combined setup is shown in Fig. 1. For aligning and calibrating the setup a stable reference beam (LO beam) was included as well as a calibrated quadrant photo diode pair (CQP) [4] on the optical bench. The Telescope Simulator has also a phase camera for beam diagnostics and an additional Gaussian beam for comparison and a back-up for the flat-top beam. The Telescope Simulator base plate has a Zerodur® tip-tilt mount for alignment so that the whole setup has an ultra low thermal expansion coefficient. Pictures of the setup are shown in Fig. 2 and Fig. 3. All optics are bonded using hydroxide catalysis bonding [5] and precision alignment techniques similar to those developed for the LISA Pathfinder optical bench interferometer [6].

Two different designs of imaging systems will be tested and compared. One is a classic four lens imaging optic, the second one has only two lenses (see Fig. 3). The tilt-to-pathlength

Figure 2. Photograph of the Optical Bench with the Telescope Simulator on top. The beams are going through a hole in the Telescope Simulator base plate to the Optical Bench via periscope optics. The baseplate of the Optical Bench is 58 cm in diameter.

coupling is measured by tilting the RX beam via two piezo driven mirror mounts and measure coupling factors in the science interferometer. The requirement is to reduce the coupling to a few μm/rad. In addition the experiment will investigate alignment tolerances of the lenses and compare the results to the optical simulations. This is important to be able to adapt the imaging systems to the exact mission parameters.

For operation the setup is placed in a vacuum chamber and operated with 1064nm laser light prepared on a three-arm modulation bench, providing three different laser frequencies with kHz heterodyne frequencies.

Figure 3. Optical bench with the two different imaging systems to test (left, photography courtesy University of Glasgow) and Telescope Simulator (right)

The construction and integration of the experiment is finished and the investigation of the tilt-to-pathlength coupling has started. The setup is a great test bed to test a lot of aspects of the eLISA interferometry. Most components have been developed for an elegant breadboard design for LISA and therefore to be space qualified. The all-Zerodur design should make the setup stable to the picometer level as required for eLISA. All components except for the ultra stable optics are replaceable for flexibility. This setup will allow experiments to test picometer interferometry, MHz heterodyne frequencies, improved imaging optics, extreme low power measurements (Picowatts) and untested new ideas like an active aperture.

3. Three back-link experiment

eLISA has two test masses and optical benches in the middle spacecraft of the triangular constellation, one for each link. This is necessary because the constellation is breathing and the angle between the links is changing. To form a Michelson-like interferometer in post processing the phase relations between the individual links are needed. An optical back-link between the optical benches is used that has to be reciprocal in both directions to single picometers. This is possible with a fiber-based link with sufficient straylight suppression [7]. To investigate alternative solutions with less straylight issues a setup with three different back-links is currently planned and built. It includes two optical benches that can be rotated with regard to each other. The tested solutions are a free beam back-link with actuated mirror mounts and a fiber based link that includes an additional laser frequency to separate the stray light by frequency.

4. Amplifier phase fidelity measurements

A possible laser source for eLISA is the laser used for LISA Pathfinder amplified by a fiber amplifier. A candidate for the amplifier is a Ytterbium doped fiber amplifier developed for inter-spacecraft communication. To test this amplifier a portable test setup is developed to measure the phase fidelity which is unknown for this amplifier. For eLISA GHz sidebands will be modulated on the laser to transmit clock information. This is necessary because the clocks of the spacecraft are not stable enough and a clock noise correction is needed. The clock noise information is imprinted on the laser by an EOM in front of the amplifier and therefor the amplifier has to transmit the sideband phase with very low phase distortions [8].

Acknowledgments

We gratefully acknowledge support by the European Space Agency (ESA) (22331/09/NL/HB, 16238/10/NL/HB) and the German Aerospace Center (DLR) (FKZ 50OQ0601, FKZ 50OQ1301) and thank the German Research Foundation for funding the Cluster of Excellence QUEST (Center for Quantum Engineering and Space-Time Research).

The University of Glasgow acknowledges the support of the the UK Space Agency (UKSA) and the Scottish Universities Physics Alliance (SUPA).

References

[1] Danzmann K *et al.* 2014 The Gravitaional Universe URL `https://elisascience.org/whitepaper`
[2] Robertson D I, Fitzsimons E D, Killow C J, Perreur-Lloyd M, Ward H, Bryant J, Cruise A M, Dixon G, Hoyland D, Smith D and Bogenstahl J 2013 *Classical and Quantum Gravity* **30** ISSN 0264-9381
[3] Wanner G 2010 *Complex optical systems in space: numerical modelling of the heterodyne interferometry of LISA pathfinder and LISA* Ph.D. thesis
[4] Fitzsimons E D, Bogenstahl J, Hough J, Killow C J, Perreur-Lloyd M, Robertson D I and Ward H 2013 *Applied Optics* **52** 2527–2530 ISSN 1559-128X
[5] Gwo D 2001 US Patent 6,284,085
[6] Killow C J, Fitzsimons E D, Hough J, Perreur-Lloyd M, Robertson D I, Rowan S and Ward H 2013 *Applied Optics* **52** 177–181
[7] Steier F, Fleddermann R, Bogenstahl J, Diekmann C, Heinzel G and Danzmann K 2009 *Classical and Quantum Gravity* **26** ISSN 0264-9381
[8] Troebs M, Bogan C, Barke S, Kuehn G, Reiche J, Heinzel G and Danzmann K 2015 *10TH INTERNATIONAL LISA SYMPOSIUM* (*Journal of Physics Conference Series* vol 610) ISSN 1742-6588

Length sensing and control for Einstein Telescope Low Frequency

Vaishali Adya[1,2], Sean Leavey[3], Harald Lück[1,2],Christian Gräf[3] and Stefan Hild[3]

[1] Max-Planck-Insitut für Gravitationsphysik (Albert-Einstein-Institut), Hannover, Germany
[2] Institut für Gravitationsphysik, Leibniz Universität, Hannover, Germany
[3] SUPA, School of Physics and Astronomy, University of Glasgow, G12 8QQ, UK

E-mail: Vaishali.Adya@aei.mpg.de

Abstract. In this paper we describe a feasible length sensing and control scheme for the low frequency interferometers of the Einstein Telescope (ET-LF) along with the techniques used to optimise several optical parameters, including the length of the recycling cavities and the modulation frequencies, using two numerical interferometer simulation packages: Optickle and Finesse. The investigations have suggested the use of certain combinations of sidebands to obtain independent information about the different degrees of freedom.

1. Introduction

The Einstein Telescope (ET) is a third-generation gravitational wave observatory with a target sensitivity of more than an order of magnitude higher than the current advanced detectors, advanced LIGO (aLIGO) [1] and advanced Virgo (AdV) [2]. ET will be located underground consisting of three nested detectors, each of which is composed of two interferometers. This xylophone configuration [3] allows the dedication of one interferometer exclusively for low frequency detection (ET-LF) and another one for high frequency detection (ET-HF). Each of the interferometers will adopt different techniques to be able to cope with different challenges. Every individual interferometer will have a dual-recycled Michelson topology with the inclusion of Fabry-Perot Arm cavities.

A control system keeps the interferometer's lengths as close as possible to the operating point in order to prevent degradation of the strain sensitivity. The signal recycling cavity of ET-LF is designed to have a detuning of 25 Hz [4]. Neither aLIGO nor AdV have been operated with detuning yet and although GEO 600 [5] was operated in a detuned mode its topology (lacking arm cavities) was different. The high finesse of the arm cavities of ET-LF results in an extremely small linear locking range thus posing a challenge to the linear control aspects. In the case of ET-LF, five degrees of freedom (DoFs) must be sensed and controlled in order to bring the interferometer to a locked state and maintain it there. These are i) differential motion of the arm cavities (DARM), ii) common motion of the arm cavities (CARM), iii) differential motion of the central Michelson interferometer (MICH), iv) length change of the power recycling cavity (PRCL), and v) length change of the signal recycling cavity (SRCL). The gravitational wave signal appears only in the DARM channel however, the auxillary DoFs must also be controlled in order to keep the interferometer at its operating point.

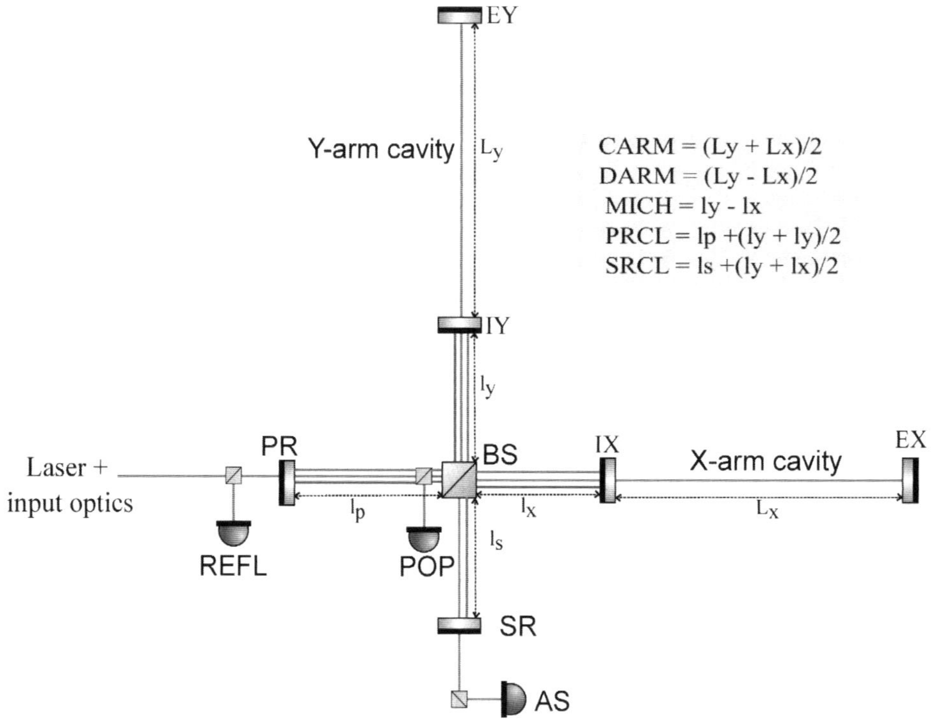

Figure 1. ET-LF interferometer design showing different sensing ports and the length DoFs. The carrier (1550 nm) is highlighted in red and the first and second RF modulation sidebands are shown in blue and pink respectively.

2. Simulation Models

In order to calculate the control matrix and the quantum noise limited sensitivity of the interferometer, an interferometer model was developed with two optical simulation softwares called Optickle [6] and Finesse [7]. Both packages have advantages of their own and the two models are in good agreement with each other in terms of the quantum noise limited sensitivity. Most of the parameters required to specify the optical system for example mirror reflectivities, masses, laser power etc. have been taken from the ET design study [4] with the exception of some lengths and tunings which we optimised using the two packages. The details of the optimisations are described in the following sections.

3. Macroscopic lengths , frequencies, DC Readout

3.1. Sideband frequencies

Sidebands are a necessary part of the control of ET-LF. The sidebands serve as additional light fields which see different resonance conditions and this is an essential trick used to decouple error signals for different DoFs. We follow a similar control scheme as the one for aLIGO, using two sidebands which are resonant in one or both recycling cavities in order to distinguish the motion of different mirrors. The first modulation frequency of 11.36 MHz (f1) is resonant in the power recycling cavity only and the second modulation frequency of 56.81 MHz (f2) is resonant in both recycling cavities. Neither of the sidebands are resonant in the arm cavities. For feasibility reasons, the modulation frequencies are limited to the range 10-100 MHz.

3.2. Macroscopic lengths and DC Readout
In order to critically couple the second sideband into the signal recycling cavity, it is necessary to introduce an asymmetry in the arm lengths of the inner Michelson interferometer called Schnupp asymmetry [8]. A compromise was made between maximising the ratio between the second and first sideband powers in the SR cavity and minimising the reflected light power from the interferometer. The power of the f2 sidebands circulating in the signal recycling cavity is maximised by choosing a small asymmetry of 8 cm which came with the added benefit of low laser frequency noise coupling. The optimal asymmetry depends on the choice of SR and PR transmissivities. With an asymmetry of 8 cm the losses for the f2 sidebands in the PR cavity to the asymmetric port are low enough to achieve high sideband gain here; yet the coupling to the SR cavity is high enough to achieve a higher power for the f2 sidebands in the SR cavity than for the f1 sidebands. The SR cavity length is set to 311.585 m in order to maintain the resonance condition for f2. A small intentional offset of about 10 pm is added to each of the end mirrors of the arm cavities to enable DC readout of the gravitational wave signal [9]. DC readout allows the gravitational wave strain to be measured without being affected by the sideband imbalance which is a consequence of the detuning introduced on the signal recycling mirror and also yields lower shot noise.

4. The Control Scheme

4.1. Sensing matrix
As a first step in designing a control scheme for ET-LF, we used an extension of the standard Pound-Drever-Hall length-sensing scheme, in which the formerly chosen radio-frequency (RF) sidebands are imposed (modulation) on the input laser beam (carrier). Following the aLIGO approach [10], we calculated the error signals generated by the beat between the sidebands for the shorter degrees of freedom. The sensing matrix presented in Table 1 indicates the DC optical gain of different DoFs at various sensing ports.

4.2. Sensing Ports/Signal Extraction Ports
There are three main detection ports that we utilised for detecting the error signals: the reflection port (REFL), pick-off port in the PRC (POP) and the anti-symmetric port (AS). REFL is the light coming back to the laser i.e. the symmetric port pick off; POP is the reflection off the AR coating of the central beam splitter; the AS port is the transmission of the signal recycling mirror. The pick off ports are designed to attenuate the power such that none of the photodiodes see more than 100 mW of power.

4.3. The matrix itself
Most of the common arm motion couples into the bright port. Hence, we would expect the CARM and PRCL signal to show up there. However, since the arm cavity finesse is large, the CARM motion dominates the PRCL motion at REFL at f1 thereby allowing us to use this signal for the CARM DoF and PRCL at POP by demodulating at the difference between the two sideband frequencies (denoted by DIFF in Table 1). All the differential arm motion couples into the dark port and hence DARM and MICH error signals show up at the AS port. We use DC readout to read out DARM, and MICH can be read out at the AS port with the demodulation frequency being the sum of f1 and f2. With DARM and MICH controlled independently, SRCL can be controlled at the AS port (even though there is a strong coupling between SRCL, MICH and DARM) by demodulating at the difference between the sideband frequencies (f2-f1). As proposed for aLIGO, we use the double demodulation technique for the short degrees of freedom (MICH, PRCL, SRCL) to derive individual error signals as these signals are not affected by cavity motion owing to the fact that the sidebands are not resonant in the arm cavities.

DoF	CARM	DARM	MICH	PRCL	SRCL
REFLQ1	**9.49e08**	1.48e06	3.03e03	-1.65e06	1.0e01
ASDC	1.55e05	**3.82e07**	7.34e05	-3.29e05	-3.62e03
ASSUMI	-1.10e03	-1.60e04	**-4.09e05**	1.51e03	-1.53e04
POPDIFFQ	-8.42e02	-2.30e01	-2.29e03	**-1.17e05**	4.89e03
ASDIFFI	6.09e03	-2.89e03	-2.06e04	-9.66e04	**-1.07e04**

Table 1. Preliminary DC Sensing matrix for detuned ET-LF obtained using FINESSE. Each row corresponds to the slope of the error signal around operating point. AS DC is the DC readout at the AS port. Other signal names consist of a signal port (REFL, POP or AS), the sideband frequency which is used for demodulation and in-phase demodulation is denoted by I and quadrature phase by Q. The preliminarily chosen ports are highlighted in bold.

5. Conclusion and Outlook

With the aid of the numerical simulation tools Finesse and Optickle, we derived a first control matrix on the way to a complete control scheme design for ET-LF. We are now working on a frequency dependent control matrix. Carrier independent error signals are obtained for the control of the shorter DoFs by double demodulation. Optimisation of the demodulation phase of the sensing photodiodes and characterisation of control loop properties in terms of stability and performance is in progress. Several other options such as amplitude modulation schemes, a single sideband approach and the use of an off-resonant sideband for the control of the shorter degrees of freedom are also under consideration.

6. Acknowledgements

This work was financially supported by the Bundesministerium für Bildung und Forschung under the ET-R&D-Networking and R&D for the Einstein Telescope project (05A2013-ASPERA-ET-R&D).

References

[1] The LIGO Scientific Collaboration et al. 2015 *Class.Quant.Grav.* **32** 074001
[2] Acernese F et al. 2015 *Class.Quant.Grav.* **32** 024001
[3] A. Freise, S. Chelkowski, S. Hild, W. Del Pozzo, A. Perreca, A. Vecchio 2009 *Class.Quant.Grav.* **26** 085012
[4] Einstein gravitational-wave Telescope 2007 Design study proposal submitted to European Union. Framework 7, INFRA-2007-2.1-01 *Design studies for research infrastructures in all S&T fields*
[5] S Hild, H Grote, J Degallaix, S Chelkowski, K Danzmann, A Freise, M Hewitson, J Hough, H Lck, M Prijatelj, K A Strain, J R Smith and B Willke 2009 *Class.Quant.Grav.* **26** 055012
[6] Evans M 2007 Optickle LIGO Document T070260
[7] Freise, A.; Heinzel, G.; Lück, H.; Schilling, R.; Willke, B. and Danzmann, K., Frequency-domain interferometer simulation with higher-order spatial modes, *Class.Quant.Grav.*, **21** , S1067-S1074 (2004). The software is available at http://www.gwoptics.org/finesse
[8] L. Schnupp, talk at European Collaboration Meeting on Interferometric Detection of Gravitational Waves (Sorrento, 1988).
[9] Fricke, T.T., et al. 2012 *Class.Quant.Grav.* **29** 065005
[10] Abbot R et al. 2010 Advanced LIGO Length Sensing and Control Final Design LIGO Document T1000298

Journal of Physics: Conference Series **716** (2016) 012031 doi:10.1088/1742-6596/716/1/012031

Early Advanced LIGO binary neutron-star sky localization and parameter estimation

C P L Berry[1], B Farr[2], W M Farr[1], C-J Haster[1], I Mandel[1], H Middleton[1], L P Singer[3], A L Urban[4], A Vecchio[1], S Vitale[5], K Cannon[6], P B Graff[7,8], C Hanna[9,10], S Mohapatra[5,11], C Pankow[4], L R Price[12], T Sidery[1] and J Veitch[1]

[1] School of Physics & Astronomy, University of Birmingham, Birmingham, B15 2TT, UK
[2] Enrico Fermi Institute, University of Chicago, Chicago, IL 60637, USA
[3] Astrophysics Science Division, NASA Goddard Space Flight Center, Greenbelt, MD 20771, USA
[4] Leonard E. Parker Center for Gravitation, Cosmology, and Astrophysics, University of Wisconsin–Milwaukee, Milwaukee, WI 53201, USA
[5] Massachusetts Institute of Technology, 185 Albany St, Cambridge, MA 02139, USA
[6] Canadian Institute for Theoretical Astrophysics, 60 St. George Street, University of Toronto, Toronto, Ontario, M5S 3H8, Canada
[7] Department of Physics, University of Maryland–College Park, College Park, MD 20742, USA
[8] Gravitational Astrophysics Lab, NASA Goddard Space Flight Center, Greenbelt, MD 20771, USA
[9] Perimeter Institute for Theoretical Physics, Ontario, N2L 2Y5, Canada
[10] Pennsylvania State University, University Park, PA 16802, USA
[11] Syracuse University, Syracuse, NY 13244, USA
[12] LIGO Laboratory, California Institute of Technology, Pasadena, CA 91125, USA

E-mail: `cplb@star.sr.bham.ac.uk`

Abstract. 2015 will see the first observations of Advanced LIGO and the start of the gravitational-wave (GW) advanced-detector era. One of the most promising sources for ground-based GW detectors are binary neutron-star (BNS) coalescences. In order to use any detections for astrophysics, we must understand the capabilities of our parameter-estimation analysis. By simulating the GWs from an astrophysically motivated population of BNSs, we examine the accuracy of parameter inferences in the early advanced-detector era. We find that sky location, which is important for electromagnetic follow-up, can be determined rapidly (~ 5 s), but that sky areas may be hundreds of square degrees. The degeneracy between component mass and spin means there is significant uncertainty for measurements of the individual masses and spins; however, the chirp mass is well measured (typically better than 0.1%).

1. Introduction

The advanced generation of ground-based gravitational-wave (GW) detectors, Advanced LIGO (aLIGO) [1] and Advanced Virgo (AdV) [2], begin operation soon: the first observing run (O1) of aLIGO is September 2015–January 2016 [3]. Binary neutron stars (BNSs) are a promising source [4].[1]

[1] Since submission, the first detection (of a binary black hole rather than a BNS), has been announced [5].

Analysis of a signal goes through several stages: detection, low-latency parameter estimation (PE), mid-latency PE and high-latency PE [6]. Each refines our understanding. To discover what we can learn about BNSs, a simulated astrophysically motivated population of BNS signals (component masses $m_{1,2} \in [1.2, 1.6]M_{\odot}$, isotropic spins with magnitudes $a_{1,2} \in [0, 0.05]$, and uniformly distributed in volume [7]) has been studied in an end-to-end analysis, with results reported in several publications. Singer *et al.* [7] studied the (low- and mid-latency) prospects for sky localization.[2] Berry *et al.* [8] repeated the analysis using more realistic noise (detector noise from the sixth science run of initial LIGO [9] recoloured to match the expected sensitivity of early aLIGO [10]), in contrast to ideal Gaussian noise. In addition to considering sky localization, Berry *et al.* [8] also investigated measurements of source distance and mass. The latter is influenced by spin, Farr *et al.* [11] completed the high-latency analysis including the effects of spin, considering all aspects of PE. We report results from these studies for O1 PE; further technical details are in the papers themselves.

2. Sky localization

Sky localization can be computed at low-latency by BAYESTAR [12] or at mid- to high-latency by LALINFERENCE [13].[3] Both are fully Bayesian PE codes; BAYESTAR uses the output of the detection pipeline, while LALINFERENCE matches GW templates to the measured detector strain [14]. Computing templates is computationally expensive; mid-latency PE is done with (non-spinning) TaylorF2 and high-latency PE is done with (fully spin-precessing) SpinTaylorT4. Both are inspiral-only post-Newtonian waveforms [15]. BAYESTAR takes a median time of 4.5 s to calculate the location [12]; the median times for the non-spinning and spinning LALINFERENCE analyses to collect 2000 posterior samples are $\sim 5.7 \times 10^4$ s [8] and $\sim 9.2 \times 10^5$ s [11] respectively.

Despite their differences, BAYESTAR and LALINFERENCE produce consistent results for a two-detector network.[4] The inclusion of spin in PE does not change sky localization for this slowly spinning population (the same may not be true for rapidly spinning black holes). At a constant signal-to-noise ratio (SNR) ϱ, there is also a negligible difference between results from Gaussian and recoloured noise. The scaling of the 50% credible region $\mathrm{CR}_{0.5}$ and 90% credible region $\mathrm{CR}_{0.9}$ with SNR is shown in Fig. 1. Assuming a detection threshold of a false alarm rate of $10^{-2}\ \mathrm{yr}^{-1}$ ($\varrho \gtrsim 10$–12), the median $\mathrm{CR}_{0.5}$ ($\mathrm{CR}_{0.9}$) is 170 deg^2 (690 deg^2) using BAYESTAR and 150 deg^2 (630 deg^2) using LALINFERENCE; switching to a threshold of $\varrho \geq 12$ [3], these become 140 deg^2 (520 deg^2) and 120 deg^2 (480 deg^2) respectively [8].

3. Mass and spin

The first estimates for the component masses $m_{1,2}$ come from the detection pipeline, here GSTLAL [16]. Full posteriors are constructed by LALINFERENCE. The degeneracy between mass and spin complicates measurements. Excluding spins (as in the mid-latency analysis) means we can miss the true parameter values. Allowing spins to vary over the full (black hole) range of $a_{1,2} \in [0, 1]$ (as in the high-latency analysis) and including precession ensures we cover the true value, but potentially means that we consider spin values not found in nature: here, the spins are $a_{1,2} < 0.05$, but we will not know the true distribution in practice.

The chirp mass $\mathcal{M} = (m_1 m_2)^{3/5}/(m_1+m_2)^{1/5}$ is the best measured mass parameter. Fig. 2(a) shows the offset between chirp-mass estimates (maximum likelihood values for GSTLAL and posterior means for LALINFERENCE) and the true values. All methods produce accurate results (offsets $< 0.5\%$) and there is no noticeable difference between recoloured and Gaussian noise. The mid-latency offsets are smaller than the high-latency ones, because our BNSs are slowly

[2] Singer *et al.* [7] also considered the second observing run (O2), with AdV joining the network.

[3] Part of the LIGO Algorithm Library (LAL) available from www.lsc-group.phys.uwm.edu/lal.

[4] This is not the case in a three-detector network if there is not a trigger from all the detectors [7, 12].

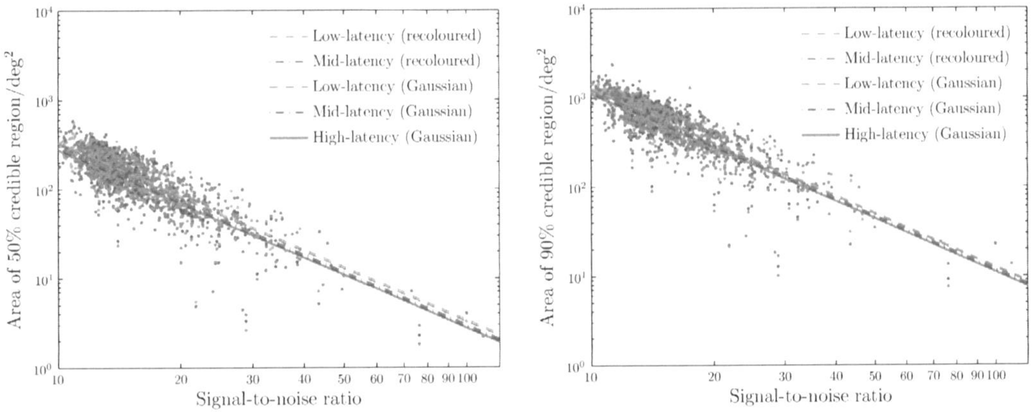

(a) 50% credible region (b) 90% credible region

Figure 1. Sky localization versus SNR for the low-latency BAYESTAR, the mid-latency (non-spinning) LALINFERENCE and the high-latency (spinning) LALINFERENCE analyses [7, 8, 11]. Individual results are indicated by points and lines indicate best fits assuming $\mathrm{CR}_p \propto \varrho^{-2}$; these are $\mathrm{CR}_{0.5} \approx (2.84 \times 10^4)\varrho^{-2}$ deg^2 and $\mathrm{CR}_{0.9} \approx (1.14 \times 10^5)\varrho^{-2}$ deg^2 across the range considered.

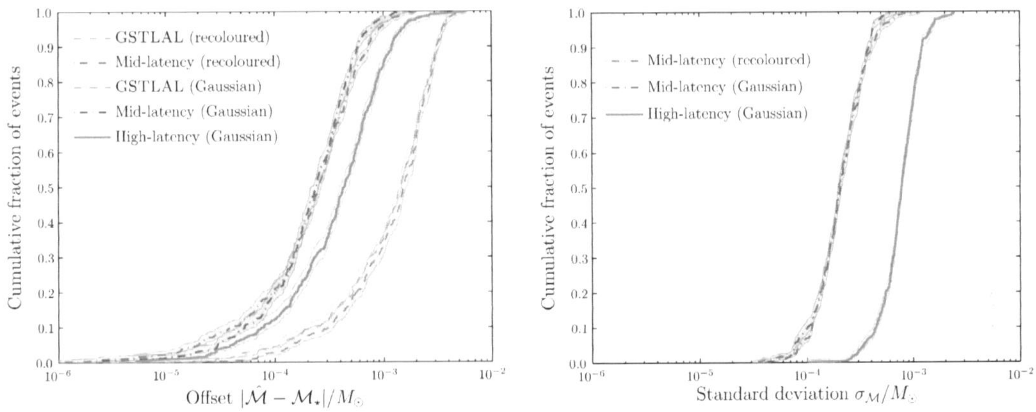

(a) Offset from true value (b) Posterior width

Figure 2. Cumulative fractions of events with (a) offsets in chirp-mass estimates and (b) posterior standard deviations smaller than the abscissa value [8, 11]. The offset is the difference between the true value $\mathcal{M}_*$ and maximum likelihood value from GSTLAL or the posterior mean from (mid- or high-latency) LALINFERENCE. The shaded areas are the 68% confidence intervals on the cumulative distributions.

spinning (which need not be the case in reality). However, the mid-latency offsets are more statistically significant. The mean values of $(\hat{\mathcal{M}} - \mathcal{M})^2/\sigma_\mathcal{M}^2$, where $\sigma_\mathcal{M}$ is the posterior standard deviation, are 5.5, 5.1 and 0.7 for the recoloured non-spinning, Gaussian non-spinning and Gaussian spinning analyses respectively. Ignoring spin yields posteriors that are too narrow [8], the distribution of $\sigma_\mathcal{M}$ is shown in Fig. 2(b) [11]; the median values of $\sigma_\mathcal{M}$ are $2.0 \times 10^{-4} M_\odot$, $2.1 \times 10^{-4} M_\odot$ and $7.7 \times 10^{-4} M_\odot$ for the recoloured non-spinning, Gaussian non-spinning and Gaussian spinning analyses respectively.

Measurements of other mass parameters, such as the mass ratio $q = m_2/m_1$ $(0 < q \leq 1)$ or

$m_{1,2}$, are less precise, and the degeneracy with spin is more pronounced [11, 14]: the median 50% (90%) credible interval for q is 0.29 (0.59). For our population of low-spin BNSs, the spins are not well measured and have large uncertainties. None of the events have a 50% upper credible bound less than 0.1; the median 50% (90%) upper credible bound is 0.30 (0.70) for a_1 (the dominant spin) and 0.42 (0.86) for a_2. Low spin values are preferred, but spin magnitudes can only be weakly constrained.

4. Summary

O1 marks the beginning of the advanced-detector era. As time progresses, sensitivities improve and further detectors (AdV, LIGO-India [17] and KAGRA [18]) come online, the prospects for detection and PE will become better [7, 19, 20]. For BNSs, chirp mass is always well measured, but sky localization and spins are more uncertain.

Acknowledgments

This work was supported in part by STFC. This is LIGO document reference LIGO-P1500155. A catalogue of results is available at www.ligo.org/scientists/first2years/.

References

[1] Aasi J *et al* (LIGO Scientific Collaboration) 2015 *Class. Quantum Grav.* **32** 074001 (*Preprint* 1411.4547)

[2] Acernese F *et al* (Virgo Collaboration) 2015 *Class. Quantum Grav.* **32** 024001 (*Preprint* 1408.3978)

[3] Abbott B P *et al* (LIGO–Virgo Collaboration) 2016 *Living Rev. Relat.* **19** 1 (*Preprint* 1304.0670)

[4] Abadie J *et al* (LIGO–Virgo Collaboration) 2010 *Class. Quantum Grav.* **27** 173001 (*Preprint* 1003.2480)

[5] Abbott B *et al* (LIGO–Virgo Collaboration) 2016 *Phys. Rev. Lett.* **116** 061102 (*Preprint* 1602.03837)

[6] Aasi J *et al* (LIGO–Virgo Collaboration) 2014 The LSC–Virgo White Paper on Gravitational Wave Searches and Astrophysics Tech. Rep. LIGO-T1400054-v7 URL `https://dcc.ligo.org/T1400054/public`

[7] Singer L P *et al* 2014 *Astrophys. J.* **795** 105 (*Preprint* 1404.5623)

[8] Berry C P L *et al* 2015 *Astrophys. J.* **804** 114 (*Preprint* 1411.6934)

[9] Aasi J *et al* (LIGO–Virgo Collaboration) 2015 *Class. Quantum Grav.* **32** 115012 (*Preprint* 1410.7764)

[10] Barsotti L and Fritschel P (LIGO Scientific Collaboration) 2012 Early aligo configurations: example scenarios toward design sensitivity Tech. Rep. LIGO-T1200307-v4 URL `https://dcc.ligo.org/LIGO-T1200307/public`

[11] Farr B *et al* 2015 (*Preprint* 1508.05336)

[12] Singer L P and Price L R 2016 *Phys. Rev.* D **93** 024013 (*Preprint* 1508.03634)

[13] Veitch J *et al* 2015 *Phys. Rev.* D **91** 042003 (*Preprint* 1409.7215)

[14] Cutler C and Flanagan E E 1994 *Phys. Rev.* D **49** 2658–2697 (*Preprint* gr-qc/9402014)

[15] Buonanno A, Iyer B, Ochsner E, Pan Y and Sathyaprakash B S 2009 *Phys. Rev.* D **80** 084043 (*Preprint* 0907.0700)

[16] Cannon K *et al* 2012 *Astrophys. J.* **748** 136 (*Preprint* 1107.2665)

[17] Iyer B, Souradeep T, Unnikrishnan C S, Dhurandhar S, Raja S and Sengupta A (IndIGO Consortium) 2011 LIGO-India LIGO Technical Report M1100296-v2 URL `https://dcc.ligo.org/LIGO-M1100296/public`

[18] Aso Y, Michimura Y, Somiya K, Ando M, Miyakawa O, Sekiguchi T, Tatsumi D and Yamamoto H (KAGRA Collaboration) 2013 *Phys. Rev.* D **88** 043007 (*Preprint* 1306.6747)

[19] Schutz B F 2011 *Class. Quantum Grav.* **28** 125023 (*Preprint* 1102.5421)

[20] Veitch J *et al* 2012 *Phys. Rev.* D **85** 104045 (*Preprint* 1201.1195)

Design study and prototype experiment of the KAGRA output mode-cleaner

Kazushiro Yano, Ayaka Kumeta and Kentaro Somiya

Department of Physics, Tokyo Institute of Technology, 2-12-1, Oh-okayama, Meguro-ku, Tokyo 152-8551, Japan

E-mail: yano@gw.phys.titech.ac.jp

Abstract. The sensitivity of the Japanese gravitational-wave detector KAGRA is limited mainly by quantum noise. In order to reduce the quantum noise level, KAGRA employs an output mode-cleaner (OMC), which filters out junk light to clean up the signal and the reference light at the signal extraction port. The proper design of the OMC is a key to achieve the target sensitivity of KAGRA. In this proceeding, we present two results. One is the final result of numerical simulations, from which we determined the optical parameters of the OMC. The other is the latest results of our prototype experiment, the goal of which is to establish the control scheme of the OMC.

1. Introduction

KAGRA is the Japanese second-generation detector based on a Michelson interferometer, with Fabry-Perot cavity in each arm (see Fig. 1) [1]. In addition, KAGRA has two optical resonators at the symmetric port and the anti-symmetric port, called the power-recycling cavity and the signal-recycling cavity. In order to maximize the observation range of gravitational waves from our primary target source *binary neutron stars*, the signal-recycling cavity length of KAGRA will be detuned from the resonance of the carrier light. KAGRA employs a homodyne readout scheme called *DC readout* [2], in which a fraction of the carrier light leaking to the dark port is used as the reference field to beat with the gravitational wave signals. The sensitivity curves and the observation ranges with different detune and readout phases have been compared to find a proper combination, which is a detune phase of 3.5 deg and the DC readout phase of 132 deg. The optimization of the readout phase is important in KAGRA where the sensitivity is limited by quantum noise.

2. Output mode-cleaner

2.1. Requirement

In this study, we assume 515 W of the laser power at the beam splitter. The radius of curvature of the test masses is 1.9 km and we assume a $\pm 1\,\%$ error between the two arm cavities. The arm cavity finesse is 1550 and the roundtrip optical loss is 100ppm. Here we also assume a $\pm 10\,\%$ loss imbalance between the two arm cavities. The roundtrip Gouy phase shifts, which describes interval of neighboring spacial modes of the carrier light, are 33 deg and 35 deg in the power and signal recycling cavities, respectively, and the incident angles at the folding mirrors in the recycling cavities are about 0.6 deg. The power reflectivity of the signal-recycling mirror is 85 %.

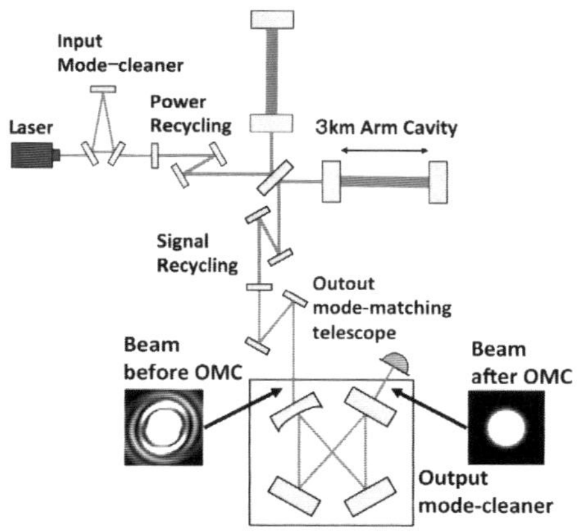

Figure 1. Schematic view of the KAGRA interferometer without mirrors for mode matching. The OMC, which consists of 4 mirrors (the first mirror is curved and the others are flat), is located at the anti-symmetric port where the gravitational-wave signal is extracted.

With these parameters, the power of the reference field at the dark port is about 1 mW. A slight differential offset on the arm cavity lengths has been applied to obtain the proper readout phase. Compared with Advanced LIGO, the room-temperature gravitational-wave detector in the United States, which does not aim at optimizing the readout phase and has the reference field of about 40 mW, KAGRA's requirement is more challenging. To realize the design sensitivity, our OMC must not degrade the shot-noise level by more than 5 % from that in the case with an ideal OMC. If the DC light power is 1 mW, a possible breakdown of an acceptable degradation after the OMC is an follows; the signal loss is to be 2% or less, the residual spatial higher order modes are to be 10μW or less, and the residual RF sidebands are to be 20μW or less [3]. This requirement shall be satisfied by selecting a proper parameter set; namely, the finesse, round-trip Gouy phase shift and length of the OMC. In order to keep the signal loss being less than 2 %, the finesse should be less than 800. It is important to select a proper combination of the Gouy phase and the cavity length, paying attention to the expected power in each higher order mode before the OMC, which should be numerically calculated using a simulation tool. In addition, we should consider the influence of the nanometer-scale surface errors of the test masses.

2.2. FINESSE simulation

In order to determine the design of the KAGRA OMC, we use a simulation tool *FINESSE*. FINESSE can analyze optical fields propagating in a user-defined interferometer not only with a plane wave approximation but also with the Hermite-Gauss modal expansion[4][5]. In the previous study of the KAGRA OMC performed by Ayaka Kumeta et al, phase maps of the test masses are introduced in the FINESSE code to simulate the expected beam distortions due to the nanometer-scale mirror surface errors [3]. A mirror phase map is an array of numbers representing the bump of the mirror surface. Since we do not have the actual mirror map of the sapphire test mass yet, those phase maps are numerically generated with the required power spectral density of the surface error taken into account.

We have modified some points of the previous simulation and obtained more accurate results. In KAGRA, the 10th mode comes out to the dark port the most since its distance to the neighboring resonance is as small as 6 deg, which is the closest among the first 19 modes. Averaging the amplitude of each mode, we provide a typically mode-distributed beam and then multiply the OMC transmittance $T_{\rm omc}$ to each mode. The results are compared with different Gouy phases and OMC cavity length. We find that $\eta = 55.4$ deg and $L = 75$ cm is the most appropriate parameter set for the KAGRA OMC. As is shown in Table. 1 and Fig. 2, the higher order modes and RF sidebands are well suppressed with this design.

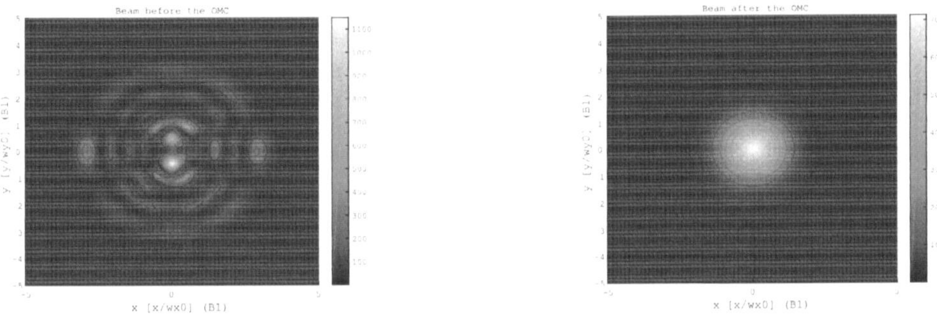

Figure 2. Beam profile of the before (left) and after (right) the OMC.

	0th	1st	2nd	3rd	10th	RF
Before OMC	1.38 mW	333 μW	6.57 mW	1.10 mW	26.1 mW	357 mW
After OMC	1.33 mW	1.96 nW	29.4 nW	76.9 nW	1.68 μW	20.4 μW

Table 1. The power in higher order modes and RF sidebands, before and after the OMC.

2.3. Prototype experiment

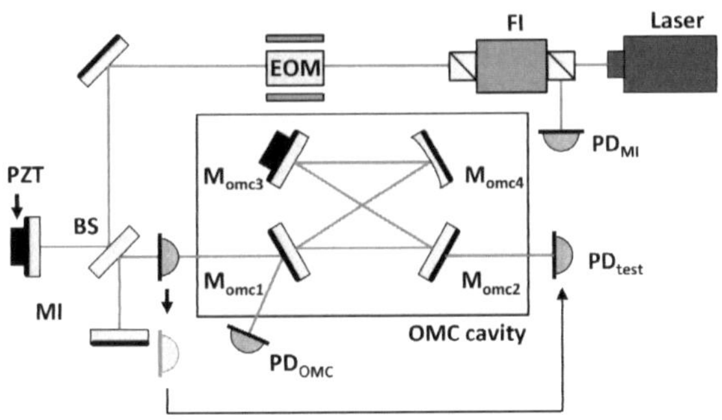

Figure 3. Schematic of the experimental layout. FI: Faraday isolator, EOM: electro-optical modulator, BS Beam splitter, MI: Michelson interferometer, PD: Photodetector. The fourth mirror (M_{omc4}) of the OMC is curved and the others are flat. PZT: piezo actuator attached.

We have constructed the prototype OMC at Tokyo Institute of Technology. The simplified experiment layout is shown in Fig. 3. The goal of this experiment is to establish the length control scheme and the parameters of this prototype are different from the actual KAGRA OMC. We use a 1064 nm Nd:YAG laser with 50 mW output power. The beam first transmits through a Faraday isolator and Electro-optical modulator that modulates the beam at 12 MHz and then is injected to a Michelson interferometer (MI). The MI is operated at its dark fringe using the Heterodyne readout technique with the reflected light extracted from the Faraday isolator (PD_{MI}) and the beam with up to 1.3 mW power is leaked to the dark port for the DC readout. The leaked beam comes into the OMC, which is a bow-tie cavity with 4 mirrors: the M_{omc4} is curved and the others are flat, the radius of curvature of the M_{omc4} is 2 m, the reflectivities $R_{omc1,4} = 0.98$, the $M_{omc2,3}$ are total reflection. We determined the cavity length to be 44 cm according to the size of our breadboard and chose the Gouy pahse of 38.7 deg and the finesse of 150 based on the calculation results using a LIGO method [6]. The OMC length is locked to the resonance of the carrier light using Pound-Drever-Hall method with the reflected light from the M_{omc1} (PD_{OMC}). With the MI being stably locked, we inject a signal at 2 kHz using a piezo actuator attached to one of the end mirrors of the MI. First we measure the output with the PD_{test} before the OMC and obtain the noise spectrum. We then measure the output with the same PD but located after the OMC and obtain the noise spectrum. Both the MI and the OMC are locked during the measurement. The left panel of Fig. 4 shows the noise spectra before and after the OMC. One can see that the amount of the signal at 2 kHz after the OMC has decreased. This is partially because the large acoustic noise of the OMC has redistributed the signal field energy. The structure around the 2 kHz signal represents the up-converted acoustic noise at lower frequencies. We know that this is due to poor mounting of the piezo actuator of the OMC and will be replaced soon. The right panel shows the ratio of the two spectra that has been normalized to have the same amount of the 2 kHz signal. The plot shows a clear improvement of the signal-to-noise ratio above 1 kHz due to the OMC.

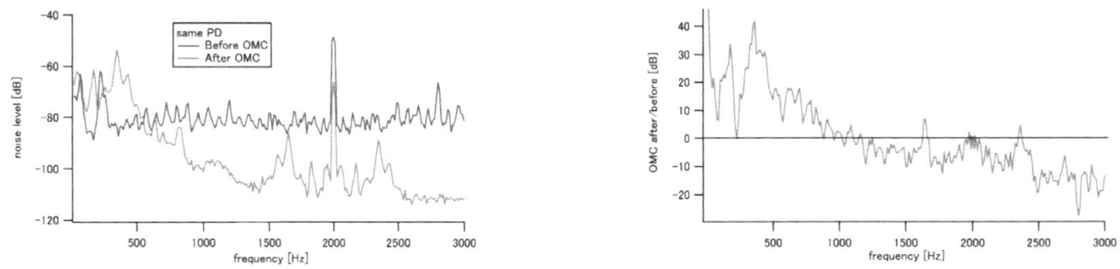

Figure 4. Left: Noise levels before and after the OMC. The amount of the signal at 2 kHz after the OMC decreases partially because of the upconverted noise around 1.7 kHz and 2.3 kHz. Right: Comparison of the signal-to-noise ratio before and after the OMC. The signal-to-noise ratio is improved above 1 kHz with the OMC. Red marker indicates the signal at 2 kHz.

3. Summary and Acknowledgement

The OMC plays an important role for KAGRA to realize the design sensitivity. The requirements of the KAGRA OMC is more challenging than other 2nd generation gravitational-wave detectors like advanced LIGO because of the low power of the DC light that is necessary to achieve the design sensitivity. In this proceedings, we modified the previous design to improve the performance of the OMC, We use a simulation tool FINESSE and numerically generated mirror maps. We also demonstrated in our prototype experiment that the signal-to-noise ratio is improved with the OMC.

This work was supported by MEXT, Leading-edge Research Infrastructure Program, JSPS Grant-in-Aid for Specially Promoted Research 26000005, and JSPS Core-to-Core Program, A. Advanced Research Networks.

References

[1] K. Somiya for the KAGRA collaboration, Class. Quantum Grav., **29**, 124007 (2012)
[2] T. Fricke, N. Smith-Lefebvre, R. Abbott, et al, Class. Quantum Grav., **29**, 065005 (2012)
[3] A. Kumeta, C. Bond and K. Somiya, Optical Review Volume22, issue1, pp149-152 (2015)
[4] A. Freise, D. Brown, and C. Bond, "*Finesse, Frequency domain INterferomEter Simulation SoftwarE*," arXiv:1306.2973 (2013)
[5] A. Freise, G. Heinzel, H. Lueck, et al, Class. Quantum Grav., **21**, S1067-S1074 (2004)
[6] S. Waldman, "Output Mode Cleaner Design" Technical Report, LIGO-T1000276 (2011)

Parameter estimation for compact binary inspirals with a simple noise realization

Jeongcho Kim[1], Chunglee Kim[2], Hyung Won Lee[1]

[1] Department of Computer Simulation, Inje University, 50834 Gimhae, Korea
[2] Yonsei University Observatory, 03722 Seoul, Korea

E-mail: jeongcho.kim@gmail.com, chunglee.kim0@gmail.com, hwlee@inje.ac.kr

Abstract. In the context of parameter estimation of gravitational waves (GWs), detector noise is assumed to be Gaussian and stationary. In reality, many electric glitches, which are neither Gaussian nor stationary, were observed and reported in publications by the LSC-Virgo collabotation. Proper noise reduction is important in GW data analysis, as these glitches would limit, if not downgrade, the quality of parameter estimation. In this work, we investigate the accuracy of results obtained by Markov Chain Monte Carlo (MCMC) parameter estimation for compact binary inspirals with the LIGO-Virgo network when non-Gaussian, stationary noise is remained in data of each interferometer. Spiky, delta function-like glitches, which are stationary, do not affect correlations between parameters. However, most likely values of chirp mass and distance seem to be shifted by the specific frequencies and amplitudes of glitches.

1. Introduction

When developing a parameter estimation (PE) library for gravitational-wave (GW) signals, detector noise is assumed to be Gaussian and stationary. However, non-Gaussian noise features (e.g. electric glitches) have been observed in engineering and science runs of the GW detector network. Many works studied appropriate treatments of detector noise existing in GW data [1, 2, 3, 4, 5, 6, 7]. In this work, we perform Markov Chain Monte Carlo (MCMC) PE for non-spinning black hole (BH) - neutron star (NS) binary inspirals with a simple model for non-Gaussian noise "glitches" at given frequencies. For a simple exercise, we adapt the frequencies of glitches found during the *blind injection challenge*[1] [8] assuming they are stationary glitches. We compare MCMC PE results with and without glitches.

Table 1. Selected source parameters for the BH-NS inspiral injection used in this work.

Parameter	$M_{\rm chirp}$ ($M_{\rm sun}$)	m_1 ($M_{\rm sun}$)	m_2 ($M_{\rm sun}$)	symmetric mass ratio	m_1/m_2	R.A. (rad)	dec. (rad)	distance (Mpc)	SNR
Value	2.994	10.0	1.4	0.108	7.1	6.485	5.747	310	19.9

[1] This corresponds to the LIGO's fifth science run and Virgo's first science run.

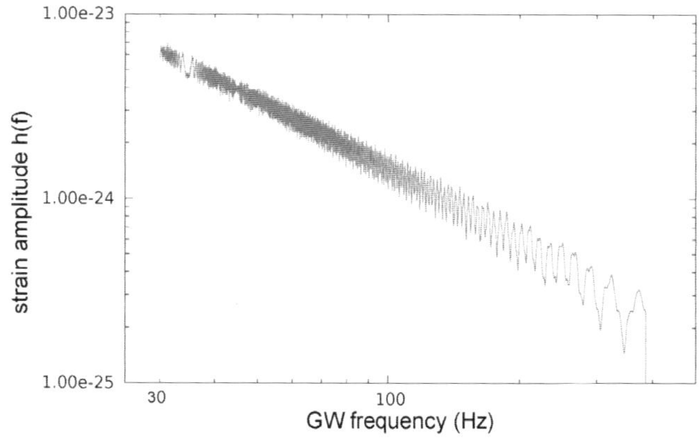

Figure 1. The injection GW signal is a full waveform generated for BH-NS inspiral in the frequency domain. We use `TaylorF2` with amplitude corrections implemented in a local branch of `LALSuite`. Note the amplitude modulations are strong due to the mass difference between BH and NS ($m_2/m_1 = 0.14$). Considering the initial LIGO-Virgo network, the injection signal is generated between the low cut-off frequency (30Hz) and ISCO frequency for a BH-NS binary (372Hz).

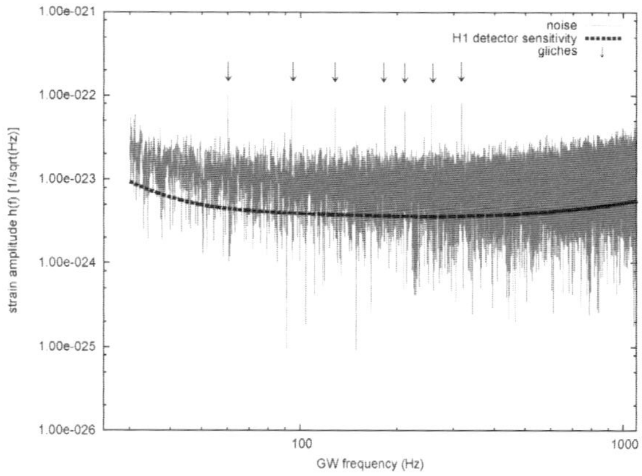

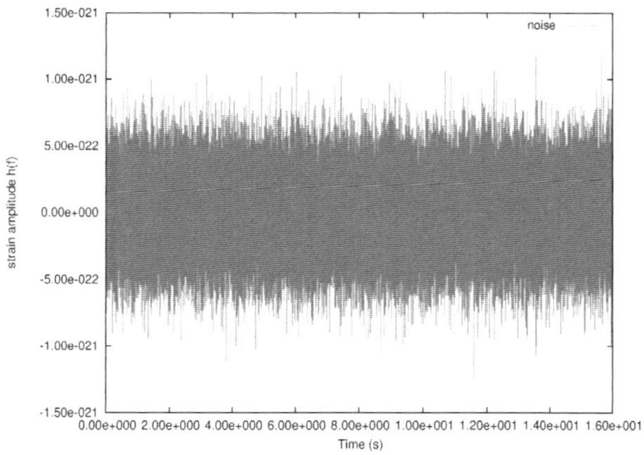

Figure 2. (a) Simulated noise (in red) for the LIGO Hanford (H1), where seven delta function-like glitches are injected to the Gaussian noise realization. We overlay the strain amplitude spectral density for H1 in a dotted black line for comparison. Black arrows indicate frequencies of artificial glitches. Noises for the Livingston (L1) and Virgo (V1) detectors are similar to that of H1. (b) We present the inverse Fourier transformed noise (from the left panel) in the time domain.

2. Generating a data

For this work, we use the `LIGO Algorithm Library (LALSuite)` in order to create a mock data, i.e. data(f) = signal(f) + noise(f) and use `LALInference` to perform MCMC PE. We consider a BH-NS inspiral as an injection, assuming a signal-to-noise ratio of 19.9. The inspiral length is roughly 12 s, between the low cut-off frequency of 30 Hz and the innermost stable circular orbit (ISCO) frequency of 372 Hz. The strain of the injection signal h(f) is shown in Fig. 1. The injection and template waveforms are generated by the frequency-domain waveform model (called TaylorF2) based on the post-Newtonian (pN) formalism. The model takes into account up to 3.5 pN order corrections for phase and 2.5 pN corrections for amplitude of a GW signal. Selected source parameters for an injection is listed in Table 1. In addition to the Gaussian noise realization for each detector, we add seven delta function-like glitches. For simplicity, we assume the frequencies of glitches are the same at all three detectors in the LIGO-Virgo network (labeled as L1, H1, and V1), where $f_{\rm glitch} = 60, 94, 128, 183, 212, 256,$ and 317 Hz. The values of glitch

Journal of Physics: Conference Series **716** (2016) 012033
doi:10.1088/1742-6596/716/1/012033

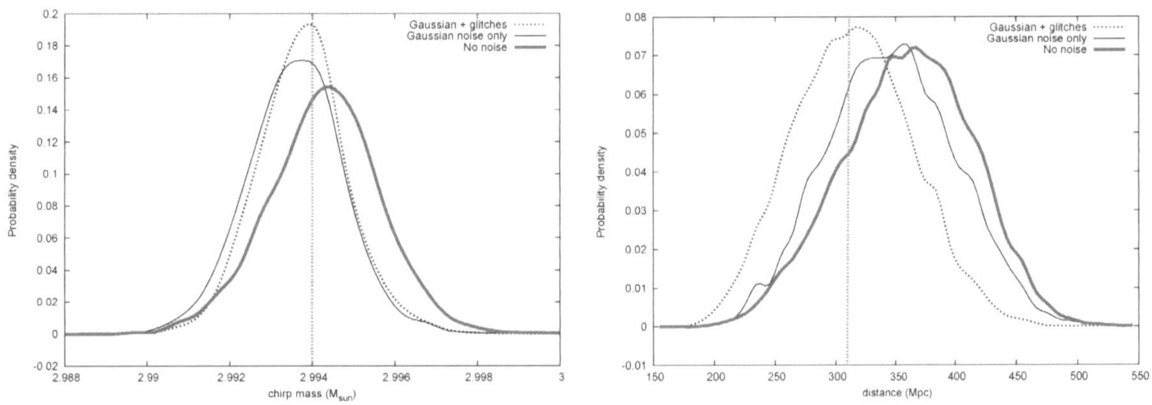

Figure 3. Posterior probability density functions (PDFs) of (a) chirp mass ($M_{\rm chirp}$) and (b) distance.y. We present results from three noise models: no noise (red thick solid lines), Gaussian noise only (thin solid lines), Gaussian + glitches (blue dotted lines). The green dotted vertical lines indicate the injection parameters.

frequencies are directly read by eyes among strongest glitches appeared in [8] (the LIGO blind injection experiment). A glitch amplitude is chosen by amplitude $= 10 \times \frac{1}{2}[PSD(f)]^{\frac{1}{2}}$, where PSD stands for a power spectral density (see Fig. 2 for the generated Gaussian with glitches for the initial LIGO Hanford noise curve). All glitches are assumed to be stationary.

3. Results

In Figs. 3 and 4, we compare PE results obtained from three noise models: (a) no noise (red thick solid lines), (b) Gaussian noise only (black thick solid ines), and (c) Gaussian + glitches (blue dotted lines). We present standard deviations of source parametes obtained from our reference MCMC run in Table 2. As one can see from these results, starionary, spiky glitches affect the most likely values and overall shape (width) of the posterior probability density function (PDF). However, standard deviations are not changed significantly.

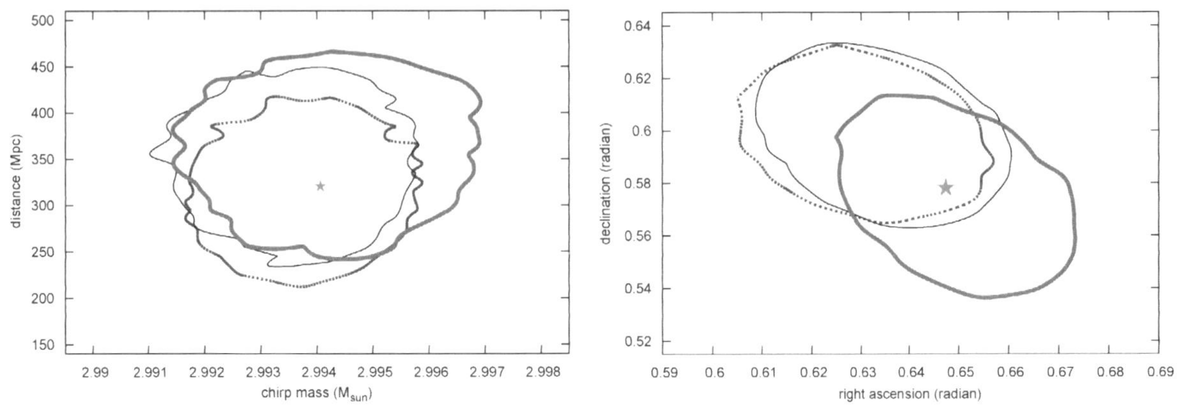

Figure 4. 2-D correlation plot of (a) chirp mass - distance and (b) right ascension - declination of the source in the Equatorial coordinate. Line color and style are the same with Fig. 3. Green stars indicate the injection values. The correlations are roughly consistent with all models, although the no noise model is least biased with respect to the injection parameters.

4. Conclusion

We examine effects of non-Gaussian glitches on PE for inspiral GW signals, assuming one of the simplest glitch model, i.e. stationary glitches with fixed frequencies like sine waves. We perform MCMC PE for non-spinning BH-NS inspirals with glitches embedded in Gaussian noise realizations. We find the medians and most likely values of individual parameters are affected by the existence of these delta function-like glitches. However, the overall widths of individual PDFs or correlations between two parameters are not significantly affected. In this work, we assume seven delta function-like glitches at given frequencies for simplicity. Realistic glitches observed from laser interferometers, however, have complicated feature in both frequency and amplitude. Moreover, many of them are not stationary. In this work, we show even the most simple form of glitches should be removed as a prerequisite of PE. We now have the machinery to inject glitches to Gaussian noise realizations for each detector using `LALsuite`.

Table 2. Standard deviations of selected source parameters obtained from different noise realizations.

	no noise	Gaussian noise only	Gaussian+glitches
chirp mass (M_{sun})	0.0014	0.0012	0.0011
distance (Mpc)	53.5833	52.8344	49.3058
right ascension (rad)	0.0114	0.0123	0.0122
declination (rad)	0.0191	0.0180	0.0170

Acknowledgments

This work is partially supported by NRF, No.NRF-2013R1A1A2060677 and by collaborative research program through KISTI/GSDC. CK is grateful for support from KASI-Yonsei DRC program of Korea Research Council of Fundamental Science and Technology (DRC-12-2-KASI).

References

[1] Allen B, Creighton J D E, Flanagan E E and Romano J D 2002 *Phys. Rev. D* **65**(12) 122002 URL `http://link.aps.org/doi/10.1103/PhysRevD.65.122002`
[2] Allen B, Creighton J D E, Flanagan E E and Romano J D 2003 *Phys. Rev. D* **67**(12) 122002 URL `http://link.aps.org/doi/10.1103/PhysRevD.67.122002`
[3] Clark J, Heng I S, Pitkin M and Woan G 2007 *Phys. Rev. D* **76**(4) 043003 URL `http://link.aps.org/doi/10.1103/PhysRevD.76.043003`
[4] Veitch J and Vecchio A 2010 *Phys. Rev. D* **81**(6) 062003 URL `http://link.aps.org/doi/10.1103/PhysRevD.81.062003`
[5] Ajith P, Hewitson M, Smith J R, Grote H, Hild S and Strain K A 2007 *Phys. Rev. D* **76**(4) 042004 URL `http://link.aps.org/doi/10.1103/PhysRevD.76.042004`
[6] Principe M and Pinto I M 2008 *Classical and Quantum Gravity* **25** 075013 URL `http://stacks.iop.org/0264-9381/25/i=7/a=075013`
[7] Littenberg T B and Cornish N J 2010 *Phys. Rev. D* **82**(10) 103007 URL `http://link.aps.org/doi/10.1103/PhysRevD.82.103007`
[8] Abadie et al. LSC-Virgo collaboration 2010 *Phys. Rev. D* **82**(10) 102001 URL `http://journals.aps.org/prd/abstract/10.1103/PhysRevD.82.102001`
[9] URL `http://www.ligo.org/news/blind-injection.php`

The cryogenic challenge: status of the KAGRA project

Raffaele Flaminio on behalf of the KAGRA collaboration

National Astronomical Observatory of Japan, Mitaka, 181-8588 Tokyo, Japan

raffaele.flaminio@nao.ac.jp

Abstract. The KAGRA project is building a gravitational wave detector inside the Kamioka mine (Japan). The detector is based on a laser interferometer with arms 3 km in length. In addition to its underground location the detector will be characterized by its mirrors made of sapphire and operated at cryogenic temperature. This paper describes the status of the construction at the site and gives an overview of the developments ongoing to prepare the cryogenic operation.

1. Introduction to KAGRA

KAGRA is a gravitational wave detector based on laser Michelson interferometer with arms 3 km in length and aiming at the direct detection of gravitational wave in the frequency band comprised between 10 Hz and a few kHz. The goal is the detection of gravitational waves emitted by astrophysical sources such as the coalescence of binaries made of neutron stars and black holes, the explosion of supernovae, the spinning and instabilities of neutron stars in the galaxy, the relic waves from the Big Bang as well as many others very energetic events undergoing in the Universe. Their detection will not only allow to probe the gravitational force in a regime inaccessible on Earth and in the solar system but also provide a new mean to observe the Universe [1].

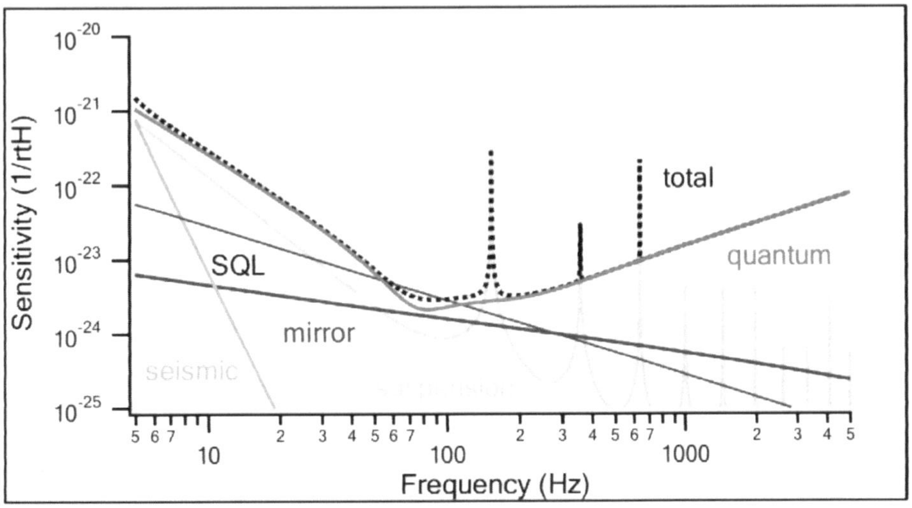

Figure 1. The KAGRA sensitivity for two different configuration of the signal recycling cavity

The KAGRA detector [2] is currently being built underground in the Kamioka mine. The site is located in the Gifu prefecture of Japan. The main features of the KAGRA design compared to other similar detectors (Virgo [3], LIGO [4], GEO[5]) is its location underground and the operation of the interferometer mirrors at cryogenic temperature. The interferometer optical configuration will be more traditional with Fabry-Perot cavities in the interferometer arms to amplify the gravitational wave effect and the use of power and signal recycling to decrease the quantum noise limitation. For the vibration isolation KAGRA will use 12 m long chains of pendulums and springs to isolate the main mirrors from the ground vibrations. The target detector sensitivity is the one shown in Figure 1 and should allow to detect the coalescence of a binary made of two neutron stars up to distances of about 150 Mpc. The overall plan foresees two phases. In a first step a simplified version of the detector (called iKAGRA) consisting of a 3 km long Michelson interferometer (without Fabry-Perot nor recycling cavities) will be operated at room temperature. Then the full interferometer (called bKAGRA) with all the components and the cryogenic mirrors will be installed and operated. The present plan is to complete the first step by the end of 2015 and the second step in 2018.

2. Status of construction

The project was approved in 2010 but, due to the big earthquake in 2011, construction could start only in 2012. The 3 km tunnels excavation including the central large hall as well as the smaller terminal halls was completed in March 2014 [6]. The excavation included the construction of four rooms above the main halls and four shafts to host the long vibration isolations systems.

After the excavation was completed the vacuum systems installation started. This includes the 3 km long vacuum tubes as well the four cryostats and all other vacuum chambers to host the interferometer mirrors. The installation was completed in the spring of 2015 despite the difficulties with the water leaking into the tunnels and the halls through the roofs and the floors. During the snow melting season the amount of water through the tunnels was more than 1000 t/h. Thanks to the 0.3% slope built in the tunnels, it was possible to avoid the flooding of the arms but additional works were required to mitigate the water issue. Nevertheless it was possible to install all the vacuum system required for iKAGRA. In addition all the clean-booths were installed around the vacuum chambers to allow installing the suspension and the mirrors in a cleaner environment.

In parallel the facility inside the cavern was completed with the construction of the clean room for the laser and the room for hosting the electronics and control system racks. Infrastructure works were done also at the exterior. In particular a building was realized outside to host the control room and some office space. The network was installed in the tunnel so to control the equipment inside the tunnel from the control room.

Thanks to the installation of the first clean booths, after the water issue was solved, it was possible to resume the installation of the equipment inside the vacuum systems. At the time of writing the first vibration isolation systems for the input mode-cleaner are being installed. In parallel the 2W laser for iKAGRA and the input optics on the laser bench were installed in the laser room. The next step will consist in sending the laser light into the input mode-cleaner and start the commissioning of the input optics system. Even if the planning is extremely tight the KAGRA project is working hard to complete the installation of the iKAGRA interferometer by the end of the year and have a one month engineering run at the beginning of 2016.

3. Preparation of bKAGRA

The final configuration of the KAGRA interferometer is called bKAGRA. Its optical configuration [7] will be similar to that of Advanced LIGO [4] and Advanced Virgo [3] i.e. the so called dual recycling Michelson interferometer with Fabry-Perot cavity in the arms. As in Advanced LIGO, KAGRA will use folded recycling cavities to avoid the cavities degeneracy [8] for the laser frequency sidebands. The main difference will be the Fabry-Perot cavity finesse which will be higher in KAGRA (1500 compared to 450 in LIGO and Virgo) in order to reduce the laser power inside the power recycling cavity and so the power inside the input mirror substrates. This allows reducing the heat to be extracted to cool down the mirrors. The laser power will be around 180 W. Assuming 80 W entering the recycling cavity and a recycling gain of 10, the power stored in the Fabry-Perot arms will be around 400 kW resulting in a shot noise similar to the one of other advanced detectors.

The high power laser is under development. The current design uses two 40 W fiber laser amplifiers coherently added to obtain an 80 W beam. A power very close to 80 W was achieved and work is ongoing to stabilize its frequency and power. This beam is further amplified passing through three solid state laser amplifiers in series to obtain 180 W. A power of 210 W was achieved but work is still to be done to improve the beam shape before the development of the stabilization system can start.

The vibration isolation system will be based on a chain of pendulums and springs made of maraging steel combined in a design similar to the one used in Virgo [9] and TAMA [10]. All the main interferometer mirrors will be suspended using this type of vibration isolation with the number of filters depending on the vibration isolation required. The so called Type A vibration isolation composed of a low frequency pre-isolator, a series of five seismic filters will be used to suspend the Fabry-Perot cavities mirrors [11]. Each mirror is part of a payload including the mirror itself as well as all the additional suspending and recoil masses required to control its position. The total height being 15 m it has been decided to hang these vibration isolation systems from the second floor of the cavern and having it reaching the main experimental area thought a shafts connecting the two floors. A shorter vibration isolation system including only two filters after the pre-isolator (so called Type B) will be used to suspend the beam splitter and the signal recycling mirrors while a further simplified vibration isolation (Type Bp) without pre-isolator will be used for the recycling mirrors. Finally a simpler double pendulum assembled over bench itself sustained by three stacks of masses and springs will be used for the input and output optics components. A complete Type B was recently installed for test at TAMA to check the installation procedure (see Figure 2) [12].

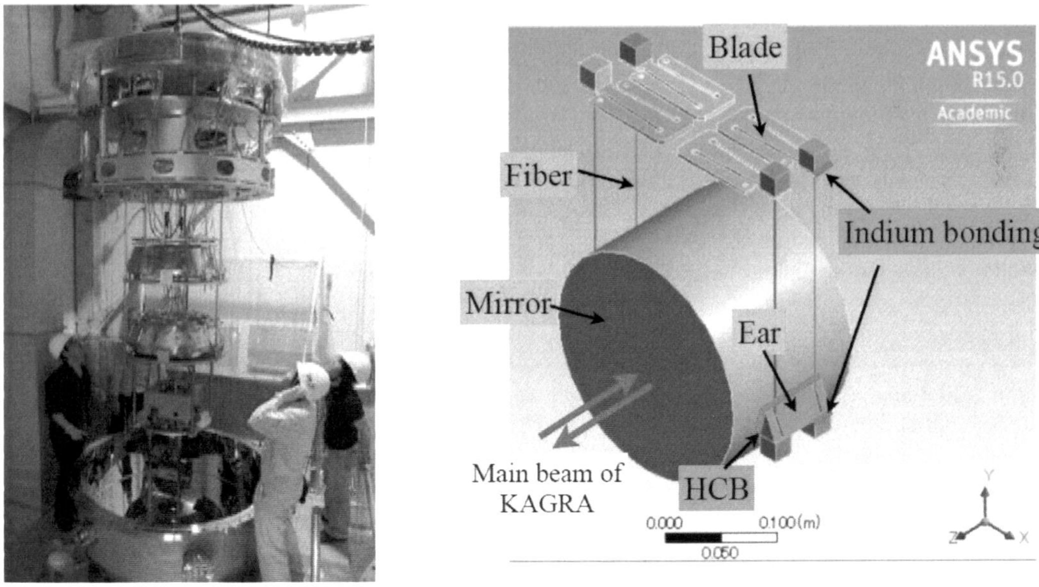

Figure 2. Test of the vibration isolation at NAOJ (left) and design of the sapphire suspension (right). HCB stays for hydroxide catalysis bonding.

The mirrors are divided into two categories. On one side the mirrors working at room temperature (including all those required for iKAGRA). These are made of silica and their production is progressing according to the schedule. On the other side the sapphire mirrors that have to be operated at cryogenic temperature. For the latter some R&D is still ongoing. The polishing on a test sample has shown that it is possible to achieve flatness rms below 1 nm as it is the case for fused silica substrates [13]. The most critical parameter is the absorption in the sapphire which is about a factor of two larger than expected (ranging from 40 ppm/cm up to 90 ppm/cm in the first two crystals received). The quality of the second two being worst, attempts are ongoing to find better crystals. If the absorption is not decreased the laser power inside the arms will have to be lowered (unless the heat extracted from the input mirrors is increased).

4. The cryogenic challenge

The cryogenic operation of the bKAGRA interferometer represents the main challenge of the project. Several aspects have to be considered. If the sapphire crystal absorption stays within the goal value of 30 ppm/cm, then 0.4 W will have to be extracted from each of the arm input mirrors. In order to do so the cryostat design includes two inner shields: an outer one at 80 K and an inner one at 8 K [14]. Four cryo-coolers will be used. The first stages of the cryo-coolers are devoted to the cooling of the outer shield. Two of the cryo-cooler second stages cool the inner shield and the other two cool the mirror payload itself. This design should allow to extract the amount of power deposited by the laser in the mirrors and cool down the mirror to 20 K provided a sufficiently heat conduction path is made between the mirror and the cryo-coolers. The temperature of 20 K has been chosen since at this temperature the sapphire used for the mirror substrate and the fibers has high quality factor. In addition two 5 m long cryogenic ducts allow the laser beam to enter the cryostat and while reducing the 300 K radiation input to the mirror by a factor of a thousand compared to the input one would have from the tube in the absence of the ducts [15].

The design of the cryogenic sapphire suspension (see Figure 2, [16]) is primarily driven by the need to cool down the mirror while keeping its thermal noise as low as possible. This is a complex system of requirements since, while the thermal noise requires to go for a monolithic suspensions all made of sapphire with the fibers suspending the mirror as thin as possible, the cryogenic requirement pushes for fibers as thick as possible. The compromise has been made to choose fibers made of sapphire 35 cm long and 1.6 mm in diameter with a nail head at both ends [17]. The lower fiber heads are attached to sapphire prisms (called ears) themselves attached to the side of the mirrors by hydroxide catalysis bonding (HCB). On their upper end the fiber heads are bonded to sapphire springs attached to the suspension upper stage (so called intermediate mass). On the ends the heads are bonded by means of a thin Indium intermediate layer. Extensive studies were performed to verify the strength, thermal conductivity and mechanical quality factor of the bondings. The mechanical quality factors of the fibers alone (not bonded to the mirror) has been measured finding values as high as 10^7 at 20 K [16] which allow reaching the thermal noise requirements. The thermal conductivity of these fibers was studied also finding values larger than 5500 W/m/K [17] and so sufficient to extract the heat provided the diameter of 1.6 mm is used. A complete prototype of the monolithic sapphire payload is currently being built [18].

The cooling time is also an important parameter since the mirrors will have to be heated up and cooled down once a year to remove the water deposited on the optical surface due to adsorption. Radiation cooling will allow to reach 150 K. Below 150 K the efficiency of radiation cooling becomes insufficient and appropriate heat links will ensure the cooling down. Moreover it is planned to coat all the payload (with the exception of the mirror, of course) with diamond like carbon (DLC). According to the simulations this solution allows increasing the radiation cooling efficiency and thus to reduce the total cooling time from nearly two months to 39 days. Experimental tests done with a half size prototype of the payload coated with DLC confirmed the validity of the simulation [15].

The heat links are a potential path for the transmission of the cryostat vibrations. This is one of the main challenge whenever cryostats are used in conjunction with an apparatus requiring a very low level of vibrations. To study this effect the cryostat vibration have been measured at the factory and combined with the expected seismic noise at the site (which is considerably lower than the one at the factory). The estimate of the cryostat vibrations is then used as input in a mechanical model of the heat links combined with a model of the suspended payload to estimate the mirror vibrations. The simulations show that even if the heat links are connected to the recoil mass of the intermediate mass, and so not directly connected to the mirror, several of the cryostat vibration peaks can limit the interferometer displacement noise. The model shows that this effect can be reduced below the noise requirement if the heat links go through another mass seismically isolated independently from the main mirror. The effect of this mass is to short circuit the heat links vibrations [19]. New vibration measurements are being done on the cryostats installed underground in the cavern to verify the validity of the previous results.

5. Data analysis preparation and perspectives

In order to prepare the scientific exploitation of KAGRA, the project is currently developing the data analysis pipelines.

A library including the tools for the data access and the data processing, called KAGRA Algorithmic Library (KAGALI) is being developed to this purpose. In parallel data analysis pipelines to be used for the search of coalescing binaries, burst event and continuous waves are being developed. In addition to a traditional matched filtering approach, the development of the coalescing binary search includes also the development of a low latency pipeline for on line search and of an algorithm based on a Bayesian approach for the estimate of the parameter of the selected events. In the case of continuous wave search a pipeline is being developed to search for signals from known pulsars using the resampling technique and GPGPU (General-Purpose computing on Graphics Processing Units). The search of burst events is being prepared by developing a pipeline based on a combination algorithms able to run both off-line and on-line.

In parallel the KAGRA project has established a collaboration with astronomers working on different type of astronomical observatories in order to prepare for the electro-magnetic follow up of candidates GW events. This collaboration, specifically funded by a grant-in-aid from the Japanese Society for the Promotion of Sciences, includes astronomers working with optical and infrared telescopes as well as radio telescope and x-ray/g-ray detectors. Last but not least the KAGRA project is preparing the joint data analysis with LIGO and Virgo. To this purpose a Memorandum of Understanding was put in place already. The final goal will be to jointly analyze the data from the three projects in order to pinpoint candidate GW sources in the sky with a precision of a few degrees and open the era of gravitational wave astronomy. Such a pointing precision will allow following up the event with other observatories and the start of a full multi-messenger astronomy including the gravitational wave sky.

This work was supported by MEXT, Leading-edge Research Infrastructure Program, JSPS Grant-in-Aid for Specially Promoted Research 26000005, and JSPS Core-to-Core Program, A. Advanced Research Networks.

References

[1] Cutler C and Thorne K S 2002 arXiv:gr-qc/0204090v1
[2] Somiya K 2012 Classical and Quantum Gravity **29** 12 SI 124007
[3] Accadia T et al 2012 Jounal of Instrumentation **7** P03012
F. Acernese et al., Classical and Quantum Gravity 32, 024001 (2015).
[4] Abbott B P 2009 Reports on Progress in Physics **72** 7 076901
The LIGO Scientific Collaboration, Classical and Quantum Gravity 32, 074001 (2015).
[5] Luck H et al 2006 Classical and Quantum Gravity **23** S71-8
[6] Uchiyama et al 2104 Classical and Quantum Gravity **31** 22 224005
[7] Aso Y et al 2013 Physical Review D **88** 4 043007
[8] Muzanmil A and Mueller G 2008 Optics Express **23** 6 557-565
[9] Braccini S et al 2005 Astroparticle Physics **23** 6 557-565
[10] Marka S et al 2002 Classical and Quantum Gravity **19** 7 1605-1614
[11] Takahashi R et al 2015 Proceedings of the MG13 Meeting on General Relativity, Eds. R. T. Jantzen and K. Rosquist 2037-2039

[12] Fujii Y et al 2015 this issue
[13] Hirose E et al 2014 Physics Review D **89** 6 062003
[14] Sakakibara Y et al 2014 Classical and Quantum Gravity **31** 22 224003
[15] Sakakibara Y et al 2015 Classical and Quantum Gravity **32** 12 155011
[16] Chen D 2015 PhD thesis (The University of Tokyo)
[17] Khalaidovski A et al 2014 Classical and Quantum Gravity **31** 10 105004
[18] Kumar R et al 2015 this issue
[19] Chen D et al 2014 Classical and Quantum Gravity **31** 22 224001

Institute of Physics
Dirac House, Temple Back
Bristol BS1 6BE UK

ISSN: 1742-6588
ISBN 978-1-5108-2454-6

CURRAN ASSOCIATES INC.
proceedings.com

2nd Workshop on Tools and Resources for REAding DIfficulties (READI 2022)

Held at LREC 2022

Marseille, France
20 – 25 June 2022